Understanding Building Automation Systems

Direct Digital Control
Energy Management
Life Safety
Security/Access Control
Lighting
Building Management Programs

Reinhold A. Carlson, PE
Robert A. Di Giandomenico
Illustrated by Carl W. Linde

R.S. MEANS COMPANY, INC.
A Southam Company

CONSTRUCTION CONSULTANTS & PUBLISHERS
100 Construction Plaza
P.O. Box 800
Kingston, MA 02364-0800
(617) 585-7880

This book was edited by John Chiang, Jill Farinelli, and Mary Greene. Typesetting was supervised by Joan C. Marshman. The book and jacket were designed by Norman R. Forgit.

Printed in the United States of America

10 9 8 7 6 5 4 3 2 1

Library of Congress Cataloging in Publication Data

ISBN 0-87629-211-2

We dedicate this book to our wives and families for their support, interest, and patience during this literary venture.

Table of Contents

Foreword   xi

Part I: Capabilities and Overview of BAS   1

Chapter 1: Development of Building Automation Systems   3

   Local Control Panels   3

   Centralized Control and Monitoring Panels   4

   Centralized Multiplexing Systems   7

   Computerized Systems   7

   Distributed Processing Systems   9

   Integrated Building Systems   10

   Future Trends and Concepts   11

Chapter 2: Distributed Process Systems   15

   Direct Digital Control   15

   Energy Management Programs – HVAC   32

   Energy Management – Lighting   41

   Fire Safety Systems   43

   Building Security/Access Control   44

   Central Host Computer   44

Chapter 3: System Architecture   51

   Remote Sensors and Actuators (Level I)   51

   Microprocessor-Based Controllers (Level II)   56

   Communications Bus   59

   Central Host Personal Computer (Level III)   59

   Management Host Computer (Level IV)   61

   Integrating Other Building Systems   62

Part II: Applications   67

Chapter 4: Direct Digital Control (DDC)   69
    Ventilation Control   70
    Heating Control   73
    Cooling/Heating, Humidification/Dehumidification Control   77
    Static Pressure Control   84
    Variable Air Volume System Terminal Box Control   85
    A Typical HVAC Unit DDC System   90
    Host Computer   90

Chapter 5: Energy Management Programs   99
    Duty Cycle Program   100
    Power Demand Limiting Program   102
    Unoccupied Period Program   104
    Optimum Start-Stop Program   105
    Unoccupied Night Purge Program   107
    Enthalpy Program   107
    Load Reset Program   108
    Zero-Energy Band Program   110
    Host Computer   110

Chapter 6: Lighting Control   115
    Occupied-Unoccupied Lighting Control   116
    Lighting Level Control   116
    Host Computer   119

Chapter 7: Fire Safety Integration   121
    Basic Fire Alarm Systems   121
    Microprocessor-Based Fire Alarm System   130
    Fire Alarm System Integration   132
    Central Host Computer   134

Chapter 8: Security-Access Control Integration   137
    Building Security Systems   137
    Access Control Systems   140
    Central Host Computer   142

Chapter 9: Facility Management Programs   143
    Computerized Maintenance Management Programs   144
    Maintenance Scheduling   144
    Utilities Metering Program   147
    Tenant Energy Monitoring Program   149
    Heating/Cooling Plant Efficiency Program   150

Part III: Planning and Estimating for BAS   157

Chapter 10: System Layout and Budget Estimate   159
   System Layout   159
   Constructing a DDC System   160
   Sequence of Operation   161
   Budget Estimating   166

Part IV: System Selection Criteria   177

Chapter 11: System Selection Criteria   179
   Considerations in Selecting a Building Automation System   180
   Vendor Selection Through Bidding Process   182

BAS Symbols   185

Glossary   186

Index   191

Foreword

For nearly twenty years, I have had the opportunity to participate in the building automation systems industry as a consulting engineer who represented users, builders, contractors, and vendors. More recently I have been a part of a successful vendor team whose goal is to listen to customers. and others within the marketplace, to ensure that future products and services are compatible with needs.

Since building automation systems utilize computers and perform complex control functions, such systems have developed an air of "magic" about them. The perception of magical forces within the systems themselves and within the industry has been reinforced by the rapid changes in a technology whose time constant is significantly shorter than the expected life of the buildings in which these systems are installed.

The level of confusion in such a rapidly changing industry has peaked, and we are now ready to recognize that the application of automated control technology to buildings is a valid engineering discipline. Furthermore, we should now be ready to understand and integrate building automation systems and disciplines into the construction process as a matter of course.

These requirements all derive their impetus from users whose expectations regarding the need for control centralization have matured along with their building operations expertise.

This book, authored by technically competent individuals who also exhibit a strong historical perspective and an end-user orientation, takes stock of current technology and industry practices. It is both an excellent text and an important reference tool. It is broad in scope, covering monitoring and control of fire safety, security, lighting and energy management functions, as well as traditional HVAC control. It is also very specific with regard to system applications, layout methodology, and estimation. Despite the historically rapid change in technology, the concepts and practices presented will not change and will serve us well for the long term.

The time has come for a generic, integrated approach to this very approachable subject. Today's state of the art begs to be documented, and we are ready to take the leap from magic and myth to competence, logic, and engineering. In my opinion, this volume will soon be a well-worn addition to the bookshelves of every design professional, contractor, builder, and building operator as we move forward in the 1990s.

Alan B. Abramson, P.E.
New York, New York

Acknowledgments

The authors wish to express their appreciation to Honeywell, Inc. for its contributions and permission to reprint selected illustrations used in this book.

Part One

Capabilities and Overview of BAS

A Building Automation System is a tool in the hands of building operations personnel, to provide more effective and efficient control over all building systems. Historically, environmental control systems for large buildings have been pneumatic. These systems are capable of maintaining acceptable environmental conditions in a building and can perform some relatively complex control sequences. However, since they are hardware-intensive, the initial installation costs and maintenance requirements can be substantial. There are also the problems of limited accuracy, mechanical wear, and inflexibility. In recent years, the integration of building systems such as heating, ventilating, and air-conditioning, lighting, fire safety, and security has proven economically advantageous, while simplifying system interaction. To determine the value of applying a Building Automation System to a particular building, one must first understand the capabilities offered by such a system. The objective of Part I of this book is to give the reader an overview of these systems, their capabilities, and the software programs associated with them.

Chapter One

Development of Building Automation Systems

The post World War II era introduced new concepts in architectural design. Radical changes in the building envelope occurred, resulting in the employment of more exterior glass, lighter construction, non-opening windows, and a strong emphasis on interior comfort conditions. While these changes enhanced the aesthetic value of the new buildings, they presented a difficult challenge for the engineer responsible for the design of the heating, ventilating, and air-conditioning system. Artificial means of introducing outdoor air for ventilation became mandatory to meet both the needs of occupants and the new ventilation codes. Mechanical refrigeration was introduced to compensate for internal heat build-up. Additional heating and cooling capacity became necessary to offset the temperature of ventilation air being introduced into the building. These requirements were satisfied by installing large, built-up, central heating, ventilating, and air-conditioning systems. The central units were installed in mechanical equipment rooms at various locations in the building where they conditioned air or water, which was ducted or piped to occupied spaces.

With these new central systems governing the building environment, automatic control became an extremely important consideration. No longer was a single, centrally located thermostat adequate to control the heat supplied to an entire building. The open window approach to cooling and ventilating buildings was a thing of the past. To meet the demand for more comfortable, well ventilated, cleaner environmental conditions in buildings, automatic temperature control manufacturers developed pneumatic and electronic control systems with year-round capabilities. Control devices were designed for increased sensitivity, and automatic control systems became more sophisticated. Environmental control engineering reached a level of major importance in the design of new buildings.

Local Control Panels

With the proliferation of mechanical systems in buildings, and the increased complexity of automatic control systems, another need became apparent. Building maintenance and operating personnel spent a considerable amount of their time walking through various mechanical equipment rooms checking the equipment, reading temperatures, and adjusting control settings. Automatic temperature control manufacturers recognized the need for centralization – both for convenience and also as a timesaving measure.

Initially, local control panels, as shown in Figure 1.1, were installed in individual mechanical rooms. These panels contained switches, pilot lights, thermometers, time clocks and appurtenances necessary to control and monitor equipment in the room. This arrangement provided a one-stop location for maintenance personnel to evaluate system status and adjust control set points.

Centralized Control and Monitoring Panels

In the early 1950s the first centralized building control centers were developed. These were freestanding, hard-wired and pneumatically piped panels. These panels contained the switches, pilot lights, and other control devices from multiple mechanical rooms. From a central location, building maintenance and operating personnel could start and stop equipment throughout the building, read temperatures at various locations, and make control set point adjustments. Graphic displays of floor layouts, mechanical systems, or piping schematics were mounted on the panel face. This gave the operator an indication of temperature sensor locations and mechanical system layout. A typical control center and sample graphic displays are shown in Figures 1.2 and 1.3.

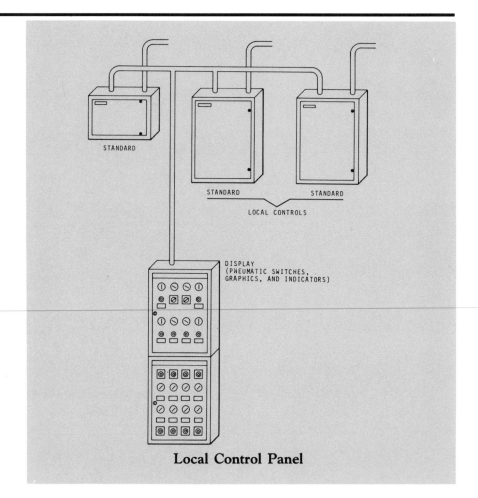

Local Control Panel

Figure 1.1

Although these systems performed their required tasks, they also had significant limitations. The initial installation cost was high due to the requirement to wire and pneumatically pipe all functions from the remote mechanical rooms back to the central panel. Expansion capabilities were also limited due to the wiring and piping burden, and the space requirements of the massive central control center panel. The systems did, however, provide some positive benefits to building maintenance and operating personnel in the form of a means by which to monitor and control building temperatures and mechanical equipment from a central location. Maintenance personnel were no longer required to walk through buildings checking mechanical equipment rooms, and could instead put their time to more effective use performing preventive maintenance.

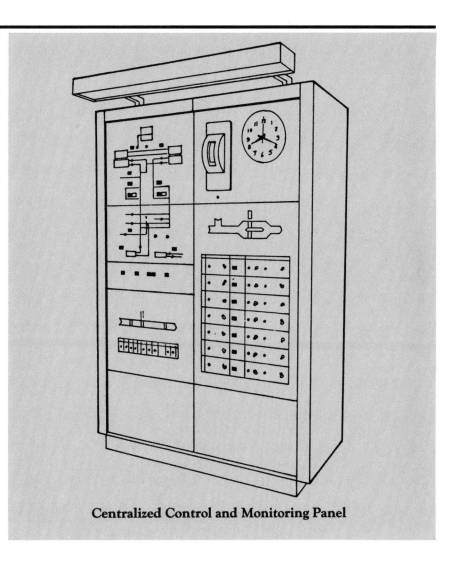

Centralized Control and Monitoring Panel

Figure 1.2

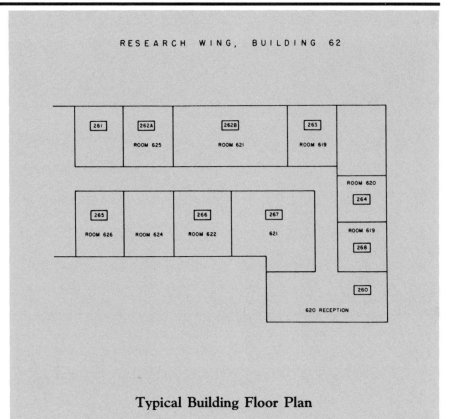

Typical Building Floor Plan

Figure 1.3a

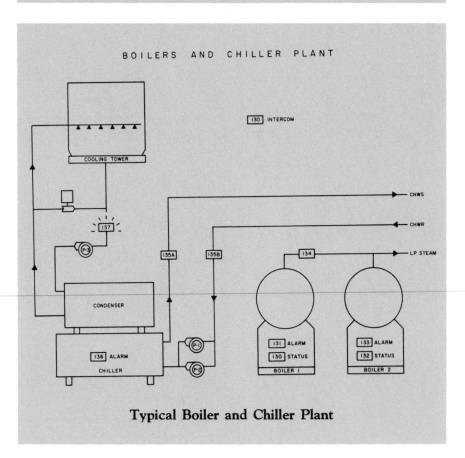

Typical Boiler and Chiller Plant

Figure 1.3b

Centralized Multiplexing Systems

In the 1960s, multiplexing was introduced to the Building Automation System. Multiplexing was a switching technology similar to that used by the telephone companies. Remote switching panels were strategically located throughout a building, and served as junction panels for temperature sensors, control set point adjustment, fan and pump start/stop wiring, etc.. These panels were connected to a central console via a multi-conductor cable. Enhancements included a slide projector and projection screen built into the console. A typical centralized multiplexing system is shown in Figure 1.4.

The projector slides could display building floor layouts; piping schematics; or heating, ventilating and air-conditioning system layouts. They could also indicate temperature sensor locations, fan and pump start/stop locations, and other pertinent data. Examples of these slides are shown in Figure 1.5. Other accessories mounted in the console included a printer to record alarms and generate run status printouts, and a recorder to document multiple temperature, humidity, or pressure points.

Computerized Systems

In the early 1970s the introduction of solid state electronics and digital transmission revolutionized the development of the Building Automation System. Computer technology using serial transmission was employed, thereby reducing the transmission loop wiring connecting the central system with remote junction panels from 80–96 conductors to a single pair. Solid state logic technology replaced noisy, electromechanical relays, and digital type sensors replaced thermocouples for temperature sensing. Data could be transmitted over transmission loop wiring at speeds up to 50 kilobits per second. The central processing unit had memory storage capability for alarm limits, time-based programming, and equipment status alarms. Digital readout of temperatures, equipment status and alarms, and printers for hard copy printout of data and alarms were featured. Operator's terminals with full color CRT displays were added.

During this period, new software packages were developed at a rapid rate. Fire and security alarm capabilities, patrol tour, card access, and lighting control software programs were included. Software programs could also be customized to the building owner's needs. These were the first "intelligent," software-driven, Building Automation Systems and the first step toward the integrated building.

The application of minicomputers to building automation and management was a major development. Increased memory capability and flexibility of system architecture made this a logical step. These computers, as shown in Figure 1.6, had the ability to perform all building automation requirements as well as generate numerous management reports. They could be installed with multiple CRT's, operator's terminals, and printers. Full English language printouts and multi-colored video displays were common. Although this technology was considered state-of-the-art, it was necessary to adapt general purpose computer software to building automation strategies. Hardware was expensive and operating personnel required a high level of training. Consequently, these systems found their application primarily in large building complexes where high paybacks on energy and manpower savings could be realized.

When the energy crisis developed in the late 1970s, the need for energy conservation in buildings became imperative. Many new programs and strategies were developed to save energy through more efficient operation of energy-consuming systems in buildings. Terms such as: *optimum start-stop, duty cycle, chiller optimization, reheat reduction,* and *load shedding* became common amongst building design and operating engineers. All of these programs were included as standard software offerings with computerized Building Automation Systems. The major automatic temperature control manufacturers embarked on development programs to enhance and expand their systems to incorporate the concepts of building and energy management into one system. A major breakthrough in cost, flexibility, and capability took place with the advent of the personal computer (PC). These systems provided a memory capacity far greater than that of any of their predecessors. They were also user-friendly, providing menu-driven selection of system graphics and points, English language descriptors and action messages, status and historical logs. PC's also make a wealth of third party software available to building operations people. Their features include: word processing, spreadsheets, maintenance management, energy audits, and bar charting. Dynamic system graphic displays provide the operator increased detail of system status and alarm condition at a glance. Programming and reprogramming of

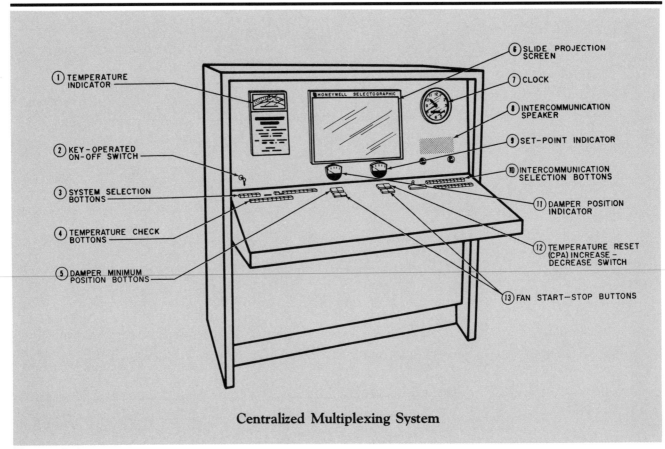

Centralized Multiplexing System

Figure 1.4

building functions can be performed by operators with little or no prior knowledge or training in computer operations. Hardware has also reached new levels of reliability and performance at reduced cost. Hard disk and diskette-drives with enhanced memory capacity, and high resolution color graphic monitors with mouse-driven pointers are featured by most system manufacturers. A great variety of high speed English language printers are also available.

Distributed Processing Systems

The introduction of the PC to the Building Automation Systems marketplace triggered the development of distributed processing systems through Direct Digital Control (DDC). Previously, the central computer was primarily an automation and energy management system with monitoring capability. The local pneumatic or electric control system provided all control functions. The remote panels, which were connected to the central computer via a two-wire transmission loop, served essentially as junction boxes for sensors and equipment control relays. The panels had no "intelligence," or automatic control capability.

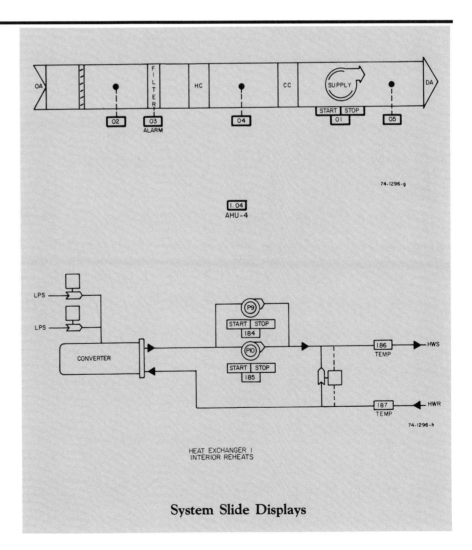

System Slide Displays

Figure 1.5

Microprocessor-based control technology, developed in the early 1980s, gave birth to "intelligent," remote panels called *direct digital control (DDC)* controllers. These are software-driven controllers which contain heating, ventilating, and air-conditioning control strategies and functions formerly performed by pneumatic or electric control hardware. They also contain all energy management programs originally resident in the central computer. These DDC controllers are wired in parallel on a two-wire communications bus to the PC. The controllers communicate with one another and with the host PC over a communications bus. The host PC can now generate management reports from information supplied to it by the network of controllers. Program changes, set point adjustments, temperature readouts, off-normal alarms, and management reports take place at the host PC. The combination of the PC, communications bus, DDC controllers, and remote sensors and actuators make up the Building Automation System of the late 1980s. A layout of this type of system is shown in Figure 1.7.

Integrated Building Systems

Other building systems, such as lighting, fire alarm, security, and access control can be readily integrated into a Building Automation System. This integration can take two forms. First, the existing, or new stand-alone fire/security system can transmit redundant alarm or status information to the local DDC controller. The host PC can display alarm data, action messages, and if desired, take action through event-initiated programs. This form of integration is usually applied to buildings where the existing fire or security system is functional.

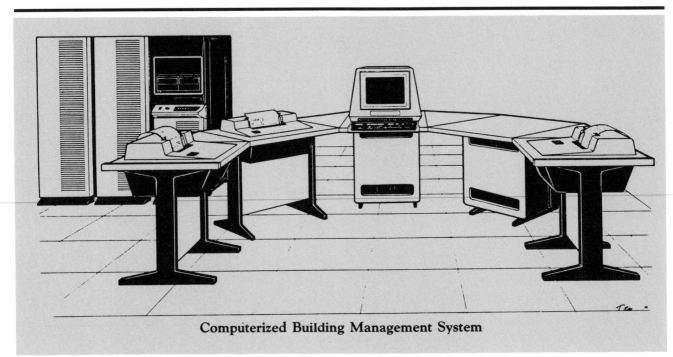

Computerized Building Management System

Figure 1.6

Under a total integrated system, the host PC directly serves all of the building systems over communication buses. The fire/security controllers can stand alone and function effectively even if the communications bus is severed or the host PC fails. Software integration and sharing of information by all controllers (HVAC, lighting, fire, security) provide a truly comprehensive mode of building control. A layout of an integrated system is shown in Figure 1.8.

Future Trends and Concepts

The Building Automation System is a tool to be used for more effective and efficient management of building operating systems. These BAS's can create optimum environmental conditions in a building while consuming minimum energy. Lighting control, fire and security protection can also be integrated into the system. The future will see continued developments in distributed processing, with the emphasis on networking systems, to facilitate communications between controllers. Systems will become even more user-friendly, with improved dynamic graphics, and window techniques where an operator can observe operations simultaneously. Software programs to

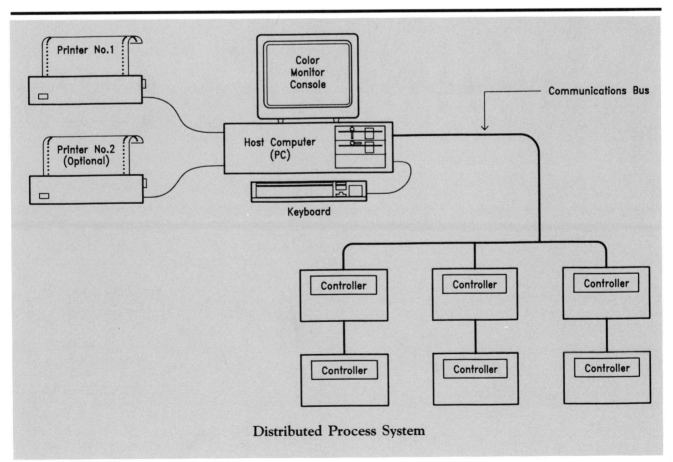

Distributed Process System

Figure 1.7

generate dynamic management reports and historical equipment information will be continually improved. This will allow building owners to constantly monitor the efficiency of various building systems. Software programs will also be developed to provide a history, on a year-by-year basis, of energy consumption in a building by building system or by floor or tenant area.

Pneumatic and electric control systems in older buildings will eventually be replaced with DDC systems. Even today, HVAC units are shipped from the manufacturer with their own attached DDC controllers. These controllers can be networked to a central PC which will control the entire building.

The concept of the intelligent building has been the subject of many technical papers in the past few years. The thought of integrating all building systems and subsystems into one comprehensive Building

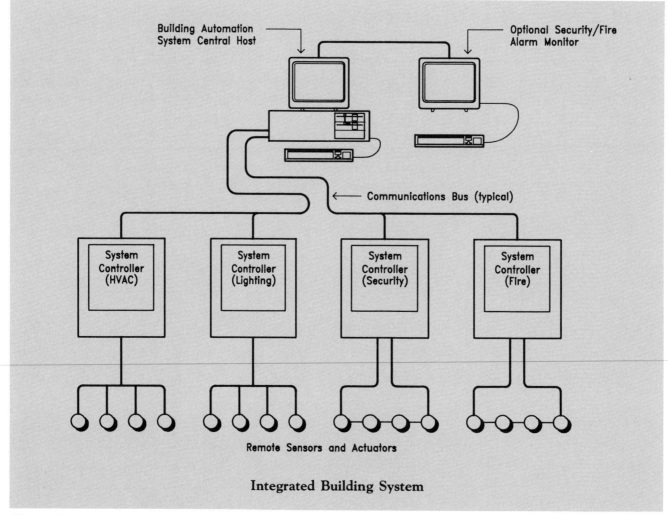

Integrated Building System

Figure 1.8

Automation System is exciting. There is no question that a building could be managed and operated more effectively and efficiently if this could be accomplished.

Historically, the integration of all building systems and subsystems under one Building Automation System has not yet occurred. One major reason is that most of these systems are manufactured, installed, and serviced by different vendors. Therefore, HVAC control systems, lighting control, fire safety systems, security and access, telecommunications, office automation, information management, and many other building systems are not compatible with one another. Some companies have taken steps to integrate some of these systems, but there are numerous compatibility questions to be addressed before in-depth integration becomes a reality. Cost is another major barrier to the intelligent building concept. Manufacturers will be required to risk large capital investments to develop a common language, or standard or open protocol, which will allow diverse digital processors to communicate with one another. However, if the demand is great enough and building owners are truly interested in system compatibility and integration, then manufacturers will undoubtedly invest the capital.

In the near future, varying degrees of systems integration will continue. This book will explore some basic steps toward integration of building systems, their control and management.

Chapter Two

Distributed Process Systems

In the original computerized Building Automation Systems all automation, energy management and control processes were performed by the central, host computer. Local control of HVAC equipment was accomplished by traditional pneumatic or electric control system. The development of microelectronic technology and the microprocessor has had a major impact on the control of all building systems, including HVAC, lighting, and life safety. These chip-sized digital computers have become comprehensive controllers and have replaced the conventional pneumatic and electric control systems. They are reliable, flexible, and provide an accuracy of control previously unachievable by traditional control systems. The processes formerly performed by the central computer are now distributed to these intelligent, microprocessor-based controllers. These include energy management programs, off-normal alarms, event-initiated programs, and time-based programs. Through a communications language, or protocol, the controllers (distributed processors) can "talk" to one another, or share data transmitted over a communications bus from a central host computer.

The communication bus can be connected to a central host computer (PC) shown in Figure 2.1. From this central computer, set points can be changed, programs revised, alarms received and documented, and management reports generated. If there is no central computer, the controllers can operate as stand-alone control systems.

Direct Digital Control

Conventional pneumatic and electric control systems consist primarily of sensors, controllers, and actuators. The sensor basically senses or measures the controlled variable (temperature, humidity, pressure, flow). The controller receives a signal from the sensor and compares this data to its set point. It then operates the actuator (damper actuator, valve actuator, pneumatic electric switch) to compensate for the disparity between set point and actual control point. This controller/actuator team can work in one of two ways: it can develop a modulating or proportioning action, which means that the actuator can assume any position in its travel, proportional to the deviation between set point and control point; or, the action can be two-position, which means that the actuator will assume one of two positions (on-off, open-closed) on a preset deviation between set point and control point. This sensor/controller/actuator arrangement is called a control loop (Figure 2.2).

In an HVAC system, these control loops occur frequently. Dampers may be controlled in a modulating or two-position mode; chilled water and hot water valves may be modulated; direct expansion cooling may be cycled in a two-position manner; VAV boxes may use both modes of control.

The controller in a Direct Digital Control (DDC) system is a digital computer. Based on the information it receives from a sensor or a switch, the controller can take any number of programmed actions, within the framework of various software programs. All of the control loops mentioned above are incorporated in the controller's software capabilities, as are many other control strategies and programs.

DDC offers numerous benefits over pneumatic and electric control systems. While electric and pneumatic control devices require periodic adjustment, calibration, and maintenance, digital computers have no moving parts and therefore require minimum maintenance. Once the sensors are in place, control loops can be changed or made interactive

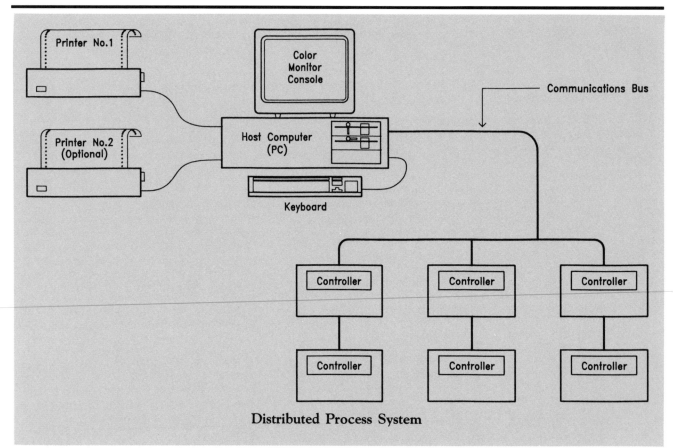

Distributed Process System

Figure 2.1

with no new wiring, piping, or devices. A centrally located PC can be linked to the DDC controllers, thereby providing the operator access to temperatures, set points, alarms, energy management programs, and other pertinent information. Accuracy of control is a feature of DDC control that is very difficult, if not impossible, to match with conventional controls.

Modulating or proportional control has an inherent problem called *offset*. Offset is defined as a sustained deviation between the set point of a control and the actual control point. With DDC control, the offset can be eliminated entirely through software programs, a difficult feat with pneumatic and electric controls.

DDC controllers have input connections to remote input devices, such as sensors or contact closure devices *(switches)*. They also have output connections to controlled devices, such as valve actuators, damper actuators, fans, pumps, etc. Input and output devices can be analog or digital. An analog input device could be a temperature, pressure, or humidity sensor which sends a continuously varying input signal to the DDC controller. A digital input is a two-position (0 and 1), on-off

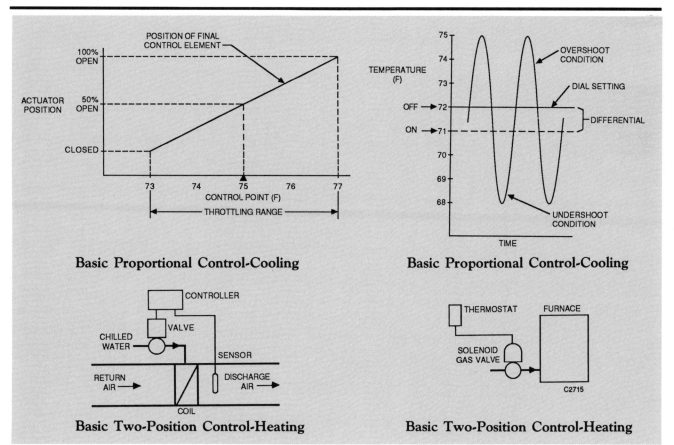

Basic Proportional Control-Cooling

Basic Proportional Control-Cooling

Basic Two-Position Control-Heating

Basic Two-Position Control-Heating

Figure 2.2

signal, such as a switch action. An analog output is a variable output which could be 0–10 volts, 4–20 MA, or 3–13 pounds for pnuematically-controlled actuators. A digital output is a two-mode output which is either on or off. This switching action could control a fan, pump, or lights.

Ventilation Control

DDC control loops used for ventilation purposes or atmospheric cooling can provide the following control sequences:

- Fixed quantity of outdoor air
- Mixed air control
- Economizer control of mixed air

Fixed Quantity of Outdoor Air

When the HVAC unit fan is energized, the outdoor air damper will go to the full open position. When the fan is de-energized, the damper will go to the closed position (Figure 2.3).

Mixed Air Control

When the unit fans are energized, the damper control system will also be energized. A temperature sensor located in the mixed air will modulate the outdoor and return air dampers to maintain mixed air temperature. A minimum outdoor air damper position can be preset to provide minimum ventilation. When the unit fans are de-energized, the outdoor air damper will fully close and the return air damper will fully open. If an exhaust damper is installed, it will modulate in unison with the outdoor air damper (Figure 2.4).

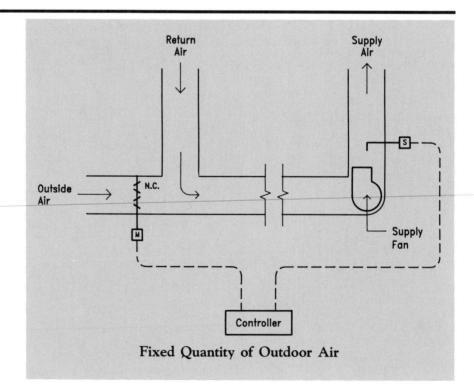

Fixed Quantity of Outdoor Air

Figure 2.3

Economizer Control of Mixed Air

When the unit fans are energized, the damper control system will also be energized. A mixed air sensor will modulate the outdoor and return air dampers to maintain mixed air temperature. A minimum outdoor air damper position can be preset to provide minimum ventilation. An outdoor air sensor will return the fresh air damper to minimum position, when outdoor air temperature exceeds a preset level. When the unit fans are de-energized, the outdoor air damper will fully close and the return air damper will fully open. If an exhaust damper is installed, it will modulate in unison with the outdoor air damper. If it is desirable to reset the mixed air temperature upward as outdoor air temperature decreases, this can be accomplished through software programming. The outdoor air and mixed air sensors are already in place to accomplish this (Figure 2.5).

Heating Control

Numerous DDC control loops and strategies are available for heating applications. Software programs are available for reset, sequencing, discharge, or space control.

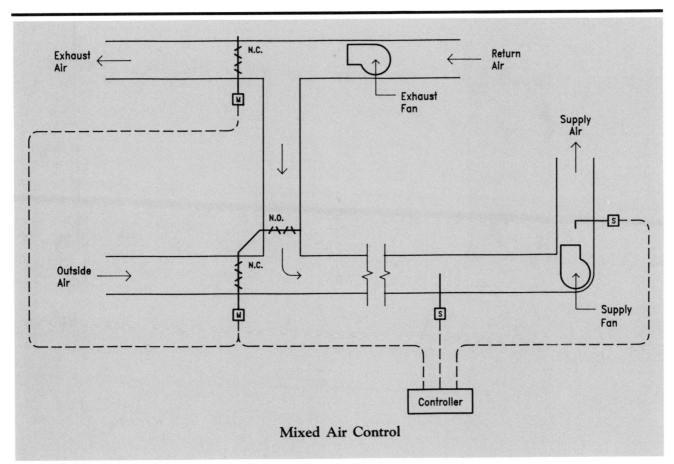

Mixed Air Control

Figure 2.4

Constant Temperature Hot Water Control (Boiler)

A sensor installed in the system supply water will modulate a three-way mixing valve to maintain its set point. The valve will mix boiler water with return water. This mixture is then pumped to the system. If the system has two-way control valves on all space heating units (radiation or heating coils), provision should be made for a differential bypass valve and sensor across the pump. As the two-way control valves close, the pressure difference across the pump will increase. The differential pressure sensor will sense this increase, and through the DDC controller, modulate the bypass valve toward the open position (Figure 2.6).

Constant Temperature Hot Water Control (Heat Exchanger)

The control loop for a heat exchanger is identical to that given above for boilers except that the three-way mixing valve will be replaced by a normally closed, two-way valve in the steam supply to the heat exchanger (Figure 2.7).

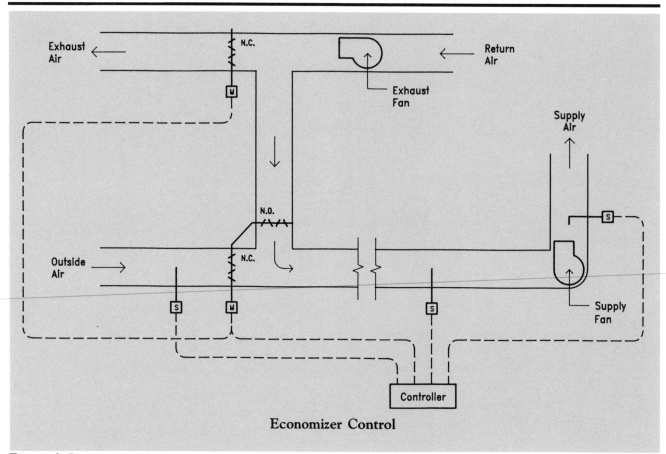

Economizer Control

Figure 2.5

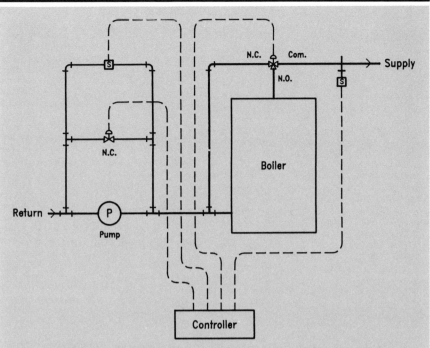

Constant Temperature Hot Water Control (Boiler)

Figure 2.6

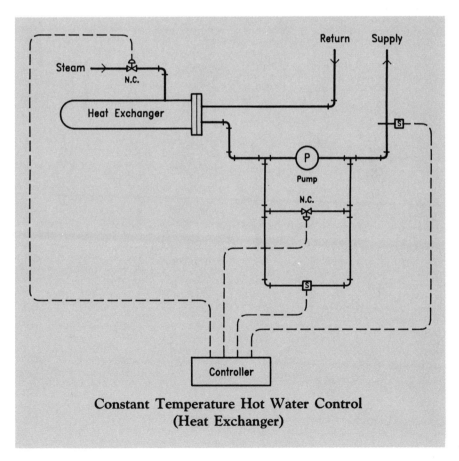

**Constant Temperature Hot Water Control
(Heat Exchanger)**

Figure 2.7

21

Hot Water Reset Control (Boiler)

The addition of an outdoor air sensor to the control loop described above for *constant temperature hot water (boiler)* will, with the help of an appropriate software program, allow reset of the discharge water temperature as a function of outdoor air temperature. A digital output point wired from the DDC controller to the pump magnetic starter will provide a means of shutting down the pump at a predetermined outdoor air temperature, when heat is no longer required (Figure 2.8).

Hot Water Reset Control (Heat Exchanger)

The addition of an outdoor air sensor to the previously described control loop for *constant temperature hot water (heat exchanger)* will allow the reset of discharge water temperature as a function of outdoor air temperature. Again, a digital output point from the DDC controller to the pump magnetic starter will provide a means of shutting down the pump at a predetermined outdoor air temperature, when heat is no longer required (Figure 2.9).

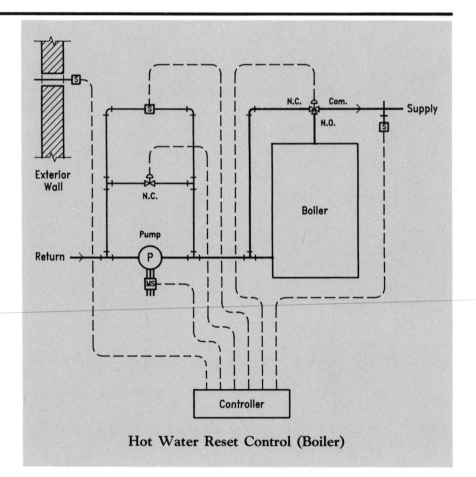

Hot Water Reset Control (Boiler)

Figure 2.8

Discharge Air Control (HVAC) – Heating

A duct-mounted sensor on the discharge side of the heating coil will modulate a control valve in the steam or hot water supply to the heating coil in order to maintain its set point. If there is any possibility that air entering the coil could be below 32°F, a low temperature safety controller should be installed on the discharge side of the heating coil. Prompted by the proper software program, the controller would initiate corrective action if a potential coil freeze-up situation should occur.

If electric heat is utilized, the same control loop is used. However, the analog output to a modulating control valve is replaced with multiple digital outputs to control stages of electric heat. If it is desirable to modulate the electric heat through a SCR (Silicon Controlled Rectifier) control, the analog output may be used in a manner similar to the modulating valve application (Figure 2.10).

Discharge Air Reset Control (HVAC) – Heating

Adding an outdoor air temperature sensor to any of the control loops described above will provide reset capability. Reset programs are available in the DDC controller to raise or lower the setting of the discharge air sensor to compensate for changes in outdoor air temperature (Figure 2.11).

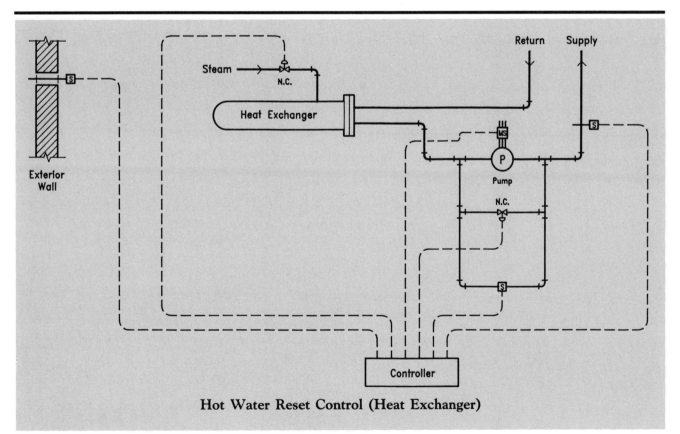

Hot Water Reset Control (Heat Exchanger)

Figure 2.9

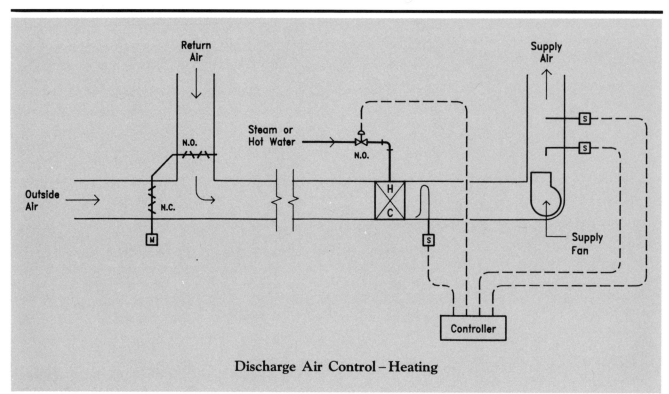

Discharge Air Control – Heating

Figure 2.10.

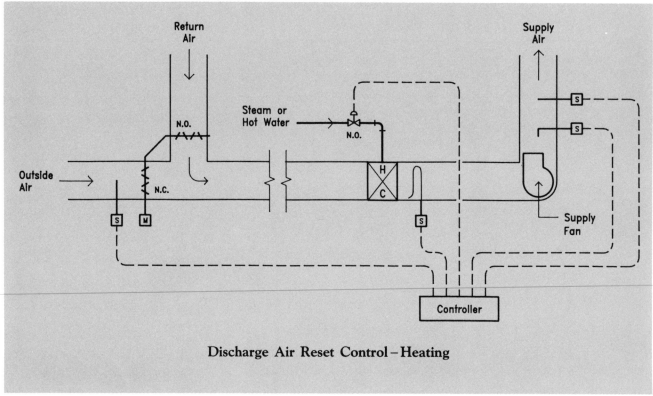

Discharge Air Reset Control – Heating

Figure 2.11

24

Space Temperature Control (HVAC) – Heating

A space temperature sensor may be installed to act as a primary controller in the applications described above. This space sensor could control the heat output of the heating coil or electric heating elements directly, with the discharge air sensor acting as a low limit. The space sensor may also reset the discharge air sensor as a function of space temperature *(Figure 2.12)*.

Cooling/Dehumidification Control

Software programs for the control of cooling and dehumidification are a component of direct digital controllers. The following are some of the programs available to control these elements:

Chilled Water Control *(Centrifugal/Reciprocating)*

A sensor installed in the chiller water supply will maintain its set point by modulating an actuator which positions the inlet vanes on a centrifugal chiller. Control of these chillers can be enhanced by installing a sensor in the return water to reset the discharge water sensor set point. Under light load conditions, the discharge water temperature can be increased to reduce the load on the chiller *(See Figure 2.13)*.

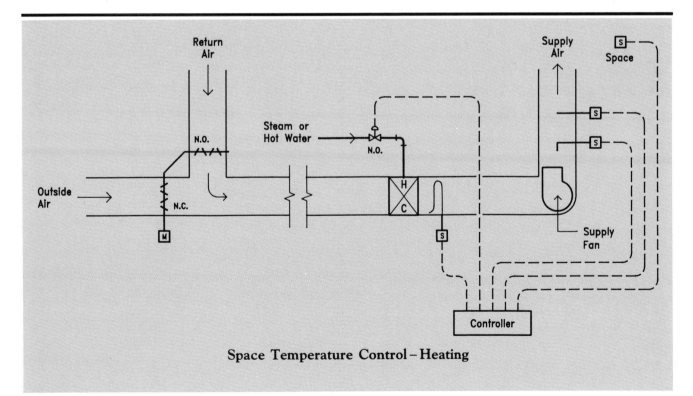

Space Temperature Control – Heating

Figure 2.12

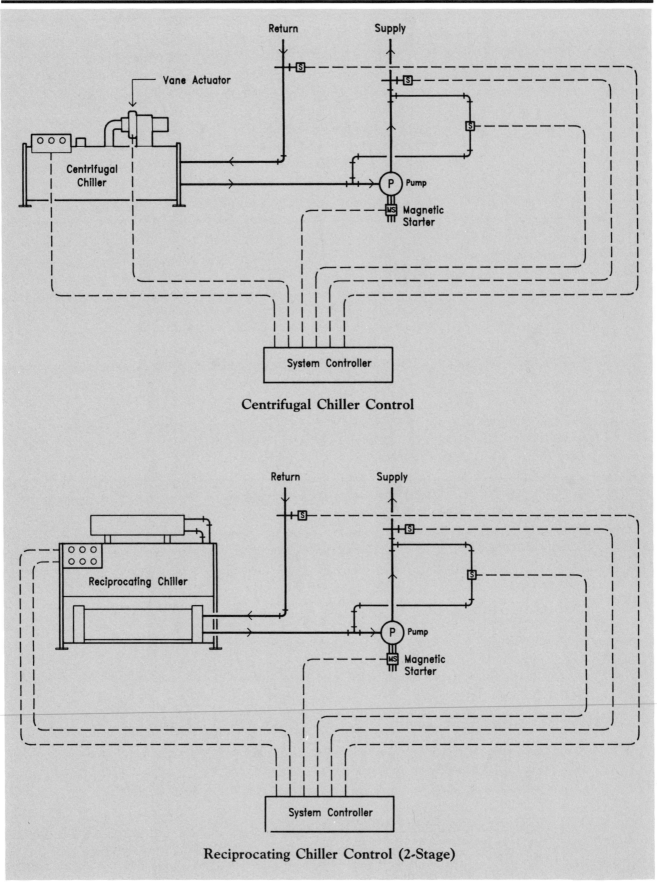

Centrifugal Chiller Control

Reciprocating Chiller Control (2-Stage)

Figure 2.13

Reciprocating chillers require digital output control rather than analog output, as in the case of the centrifugal chillers. Reciprocating chillers operate in two position stages by sequencing multiple compressors, or unloaders, as the load changes. Here again, a return water sensor can be utilized to reset the discharge water sensor as the load changes (See Figure 2.13).

Discharge Air Control (HVAC) – Cooling

A duct-mounted sensor on the discharge side of the chilled water coil will modulate a control valve in the chilled water supply to the coil, to maintain its set point. The control valve may be a three-way or two-way type. In either case, it will require an analog output from the DDC controller. (Figure 2.14).

If direct expansion (DX) cooling is used, discharge air control is not recommended. This would result in short cycling of the compressor due to the large temperature drop when the system compressor is energized. The best application would be control from a space or a return air temperature sensor.

Discharge Air Reset Control (HVAC) – Cooling

Adding an outdoor air temperature sensor to the control loops described above will provide reset capability. Reset programs are available in the DDC controller to vary the setting of the discharge air sensor to compensate for fluctuations in outdoor air temperature.

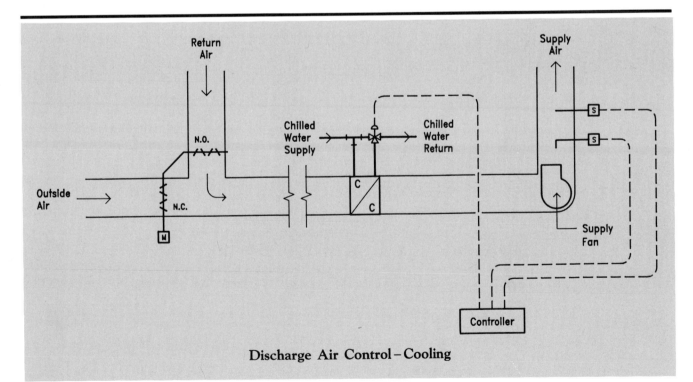

Discharge Air Control – Cooling

Figure 2.14

A space temperature sensor or multiple sensors may also be used to reset the discharge sensor set point. This will vary the discharge air temperature as a function of load in the space. In the case of multiple sensors, the area with the greatest cooling requirements will reset the discharge sensor. Averaging of sensor readings is not recommended (Figure 2.15).

Space Temperature Control – Cooling

A space temperature sensor may replace the discharge sensor to control the chilled water control valve. If direct expansion cooling is used, a space temperature sensor may be applied to cycle the compressor through a digital output point (Figure 2.16).

Dehumidification Control

Dehumidification can be accomplished by the use of a chilled water coil or a direct expansion coil. In the case of the chilled water coil, a space humidity sensor will modulate, or "two position" a control valve in the chilled water supply. A reheat coil should be provided to reheat the air after dehumidification. The space temperature sensor which controls the reheat coil may also modulate the chilled water valve for space cooling requirements. If direct expansion dehumidification is used, the space humidity sensor will cycle the compressor in a two-position manner. Here again, some method of reheat should be provided to maintain space comfort conditions (See Figure 2.17).

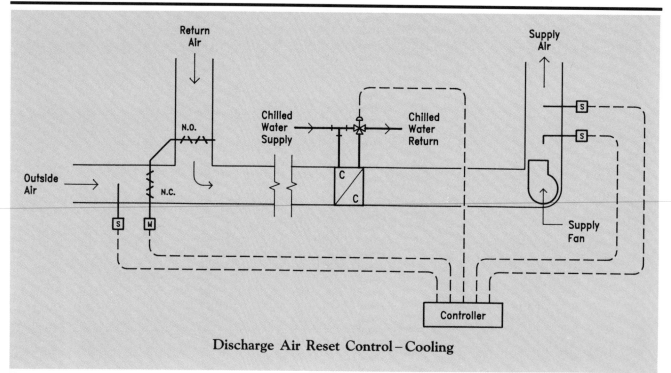

Discharge Air Reset Control – Cooling

Figure 2.15

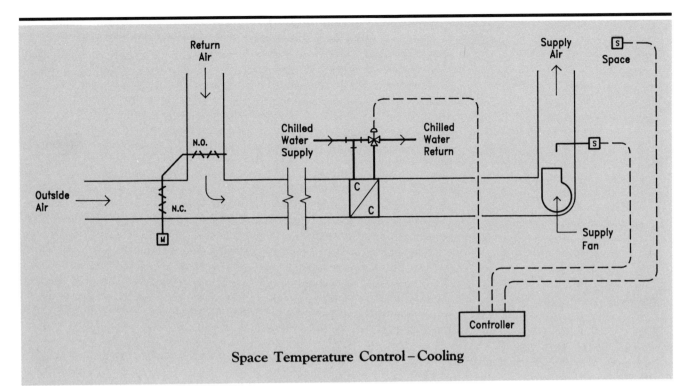

Space Temperature Control – Cooling

Figure 2.16

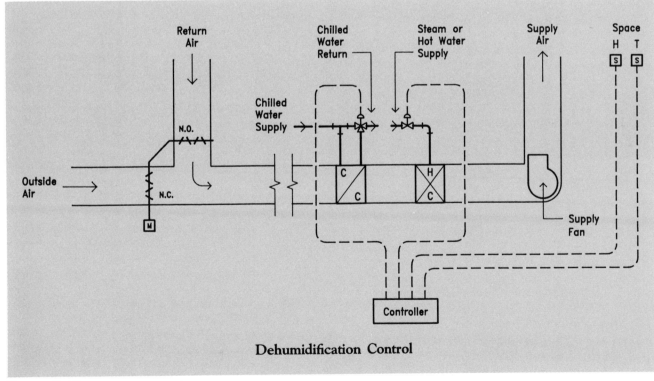

Dehumidification Control

Figure 2.17

Humidification Control

A space humidity sensor maintains space relative humidity by modulating the valve on a steam humidifier. A high limit, duct-mounted, humidity sensor located downstream of the humidifier will override the space sensor and modulate the steam valve if relative humidity in the discharge duct reaches its set point. Some humidifiers require a two-position rather than modulating control mode. In this case, a digital output point from the DDC controller is used. When the unit supply fan is de-energized, the humidifier will shut down (See Figure 2.18).

Heating/Cooling Sequencing

A single space temperature sensor, or duct-mounted, discharge air sensor may be used to control the heating coil valve and chilled water coil valve in sequence. As an energy savings feature, a deadband (or zero-energy band) may be programmed, using appropriate software, between the heating and cooling modes. During this deadband both valves are closed. Although this practice may widen the temperature swings within the space, these ranges could be limited to acceptable levels for comfort, with appreciable energy savings (See Figure 2.19).

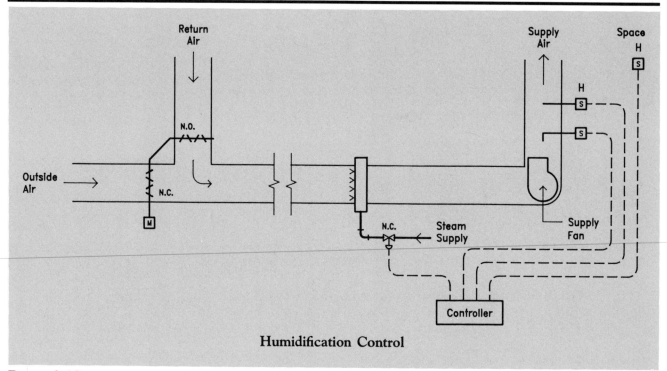

Humidification Control

Figure 2.18

Humidification/Dehumidification Sequencing

A single space humidity sensor may be used to control the humidification and dehumidification process. When space relative humidity exceeds the setting of the humidity sensor, the cooling coil valve modulates open to provide dehumidification. When relative humidity falls below the space sensor set point, the humidifier steam valve modulates toward the open position, subject to the override action of the discharge high limit humidity sensor. As mentioned under *Dehumidification Control*, a reheat coil is necessary to maintain comfort conditions during dehumidification (See Figure 2.20).

Static Pressure Control

The control of static pressure inside a building, or in variable air volume system supply ducts, can be accomplished in two ways: (1) modulating fan inlet vane dampers or (2) variable speed drives. For static pressure control in variable air volume (VAV) systems, the static pressure sensor is located in the main supply duct to the terminal units. This sensor will maintain supply duct static pressure by modulating the system fan inlet vanes (See Figure 2.21), or by modulating fan speed, through a variable speed drive. Variable frequency drives provide excellent fan speed control, and most are designed for 3-15 pound, or 4-20 milliamp inputs, which fall within the scope of most DDC

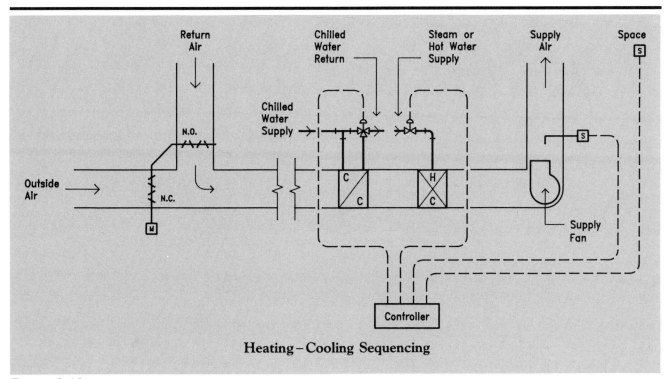

Heating – Cooling Sequencing

Figure 2.19

controllers. For direct building static pressure control, the sensor is located in the space and is referenced to a location outside the building. Space static pressure is maintained by controlling return fan inlet vanes, or by variable speed drive on the return fan.

Energy Management Programs – HVAC

In a distributed process system, the energy management programs formerly performed by the central computer, are now resident in the intelligent, microprocessor-based DDC controllers. The energy management programs described below are traditional, well-proven programs which have become standards in the industry. Most DDC systems also have the capability of custom-designed programs to operate HVAC systems in an energy efficient manner.

Duty Cycle Program

The purpose of a duty cycle program is to reduce energy consumption by periodically shutting down energy-consuming systems. Shutdown periods can be on a fixed cycle rate, or on a variable cycle rate, a choice that will depend on space and outdoor air temperature. An additional benefit of this program is its tendency to reduce electrical demand, if properly applied. Under the duty cycle program, an HVAC system is shut down for short periods of time and the space temperature is

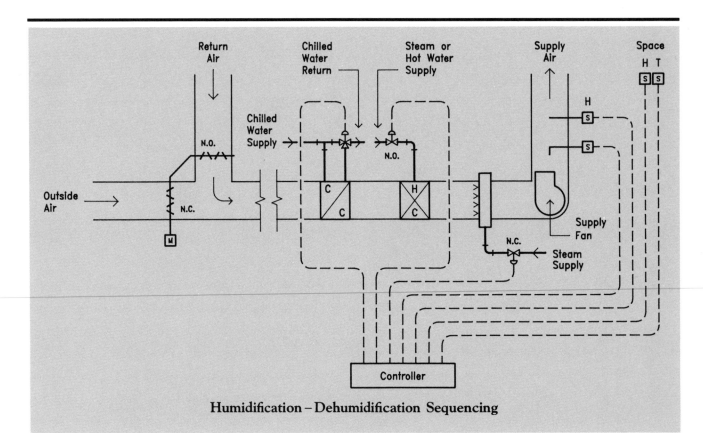

Humidification – Dehumidification Sequencing

Figure 2.20

allowed to drift within limits that do not jeopardize comfort. During the shutdown period, no fan electrical energy is consumed, and ventilation air need not be heated or cooled, since the system is not operating, and the outdoor air damper is closed.

Fixed time duty cycling means that the HVAC system will be on for fixed periods of time, and off for fixed periods of time during the duty cycle period. Different areas of a building could have different cycling requirements due to internal loads, sun loads, or building construction. With space temperature compensated duty cycling, the "on" and "off" periods are varied according to feedback from a space temperature sensor. When space temperature is within programmed comfort limits, the "off" periods are extended and the "on" periods shortened. As space comfort limits are approached, the reverse will occur. This is clearly a more effective means of achieving maximum energy savings, while maintaining optimum comfort conditions, than fixed time duty cycling.

Existing HVAC systems that have space thermostats which cycle the fan and the heating/cooling source simultaneously are not good candidates on a return on investment basis for a duty cycle program, since duty cycling is already taking place. Figure 2.22 shows a typical duty cycle schedule.

Power Demand Limiting Program
The power demand program controls electrical loads in buildings in order to eliminate demand peaks that result in costly demand charges. The controller monitors electrical demand, compares it with preprogrammed target limits, and sheds loads to limit peaks.

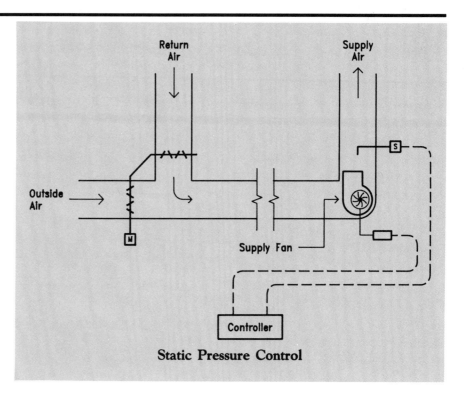

Static Pressure Control

Figure 2.21

The electrical loads in the building should be prioritized. As the building electrical demand approaches the preprogrammed limit, the lowest priority loads will be shed in sequence. Low priority loads are those that supply electric devices that the occupants can do without for short periods of time. If the demand continues to increase, the controller will begin to shed the second priority loads. If building demand continues to increase, an alarm may sound at the central computer, at which point the operator can manually shed loads from the terminal. Figure 2.23 gives an indication of how energy savings are achieved with this program.

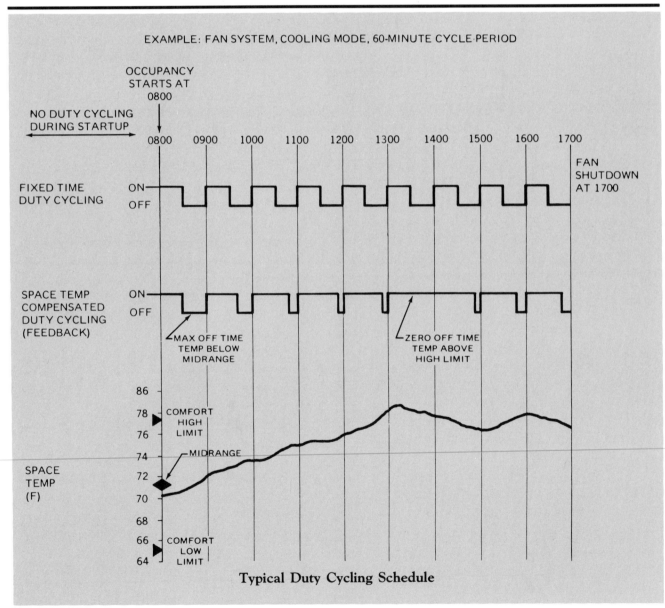

Typical Duty Cycling Schedule

Figure 2.22

Time of Day Program

The Time of Day Program is basically a time clock function which provides for shutdown of energy-consuming loads based on time only. It can shut down boilers, chillers, or HVAC units at the end of the building occupancy periods, and reactivate them prior to the next occupancy period. If this program is used to shut down heating equipment in areas where extremely low temperatures may be experienced, provision should be made to override the program in the event that space temperatures reach dangerously low levels.

Optimum Start/Stop Program

This program conserves energy while maintaining comfort when HVAC systems are started in the morning and shut down at the end of the day. Comfort levels and occupancy times are programmed into the controller. HVAC system startup prior to anticipated occupancy is delayed until the last possible moment, yet timed to have the building or zone up to programmed comfort levels at occupancy time. Shutdown of the HVAC systems takes place prior to the end of the programmed occupancy period, and the building is allowed to coast into vacancy.

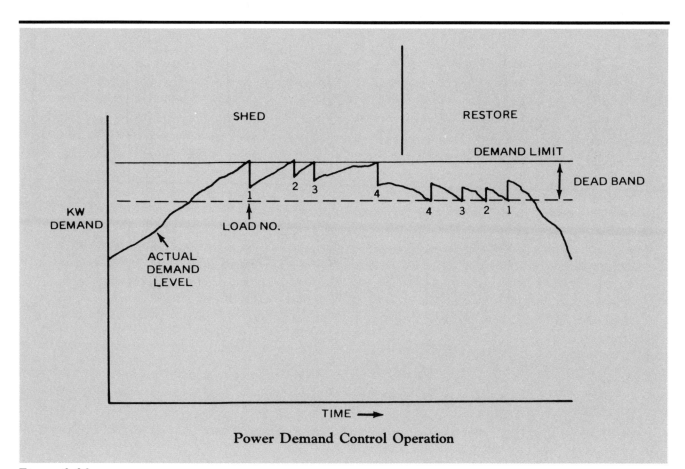

Power Demand Control Operation

Figure 2.23

The optimized startup and shutdown times take into consideration space temperature, outdoor air temperature, the preprogrammed space comfort conditions, occupancy times, and the heat loss/heat gain characteristics of the building or zone. Figure 2.24 shows some typical optimum start and optimum stop graphs.

Unoccupied Space Temperature Setback Program

The Unoccupied Space Temperature Setback Program, or night cycle program, has a twofold purpose: energy conservation and building protection. It can be applied to buildings or zones that are subject to temperature or humidity extremes, although it was developed primarily for heating applications.

This program reduces energy consumption by maintaining reduced building temperatures during unoccupied periods in the heating season. The reduced temperatures are kept within limits that will protect the contents of the building. The program can maintain outdoor air dampers on HVAC systems in the closed position whenever the fan is cycled to satisfy the reduced temperature setting. During the cooling season, the night cycle program allows building or zone temperatures to drift to higher levels during unoccupied periods. The HVAC systems may be cycled to maintain a high limit temperature or, if there is no anticipated danger to the building contents, the HVAC systems may remain in the de-energized state during the entire period. Figure 2.25 shows typical night cycle program curves for the heating seasons.

Unoccupied Night Purge Program (Summer Only)

This program can provide free atmospheric cooling shortly before the time of occupancy, thereby delaying the startup of mechanical cooling and reducing electrical consumption. A few hours prior to occupancy, the program begins monitoring indoor and outdoor conditions (temperature and dewpoint). If the need for cooling is detected, and outdoor air conditions are suitable, the program will energize the HVAC unit fan and open the outdoor air damper 100%. The exhaust fan will also be energized. This purging of the building will continue until the space temperature and dewpoint reach preprogrammed limits, at which time the fans will be de-energized. With the building pre-cooled, startup of mechanical cooling equipment and HVAC systems in preparation for occupancy can be delayed (See Figure 2.26).

Enthalpy Program

The purpose of the enthalpy program is to reduce energy consumption for cooling, by selecting either return air or outdoor air, based on which has the lowest total heat (enthalpy). The program constantly monitors outdoor air and return air temperature, and relative humidity, or dewpoint. It will then operate the mixing dampers to use outdoor air or return air, whichever has the lowest total heat. This minimizes the load on the mechanical refrigeration, thereby saving on energy (See Figure 2.27).

Load Reset Program

This program can save energy by adjusting HVAC system supply air temperature to just satisfy the zone with the greatest heating/cooling requirement. It applies to HVAC systems employing heating and cooling simultaneously. It may also be applied to chillers and hot water systems.

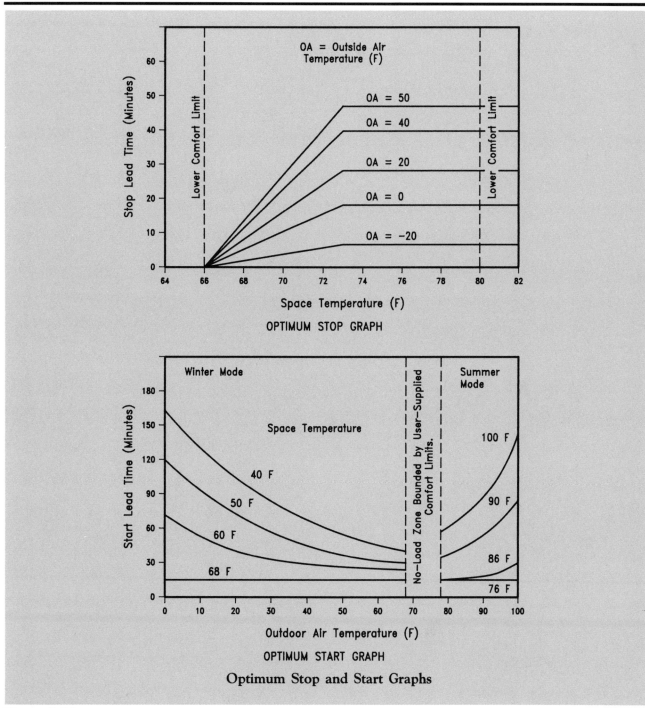

Optimum Stop and Start Graphs

Figure 2.24

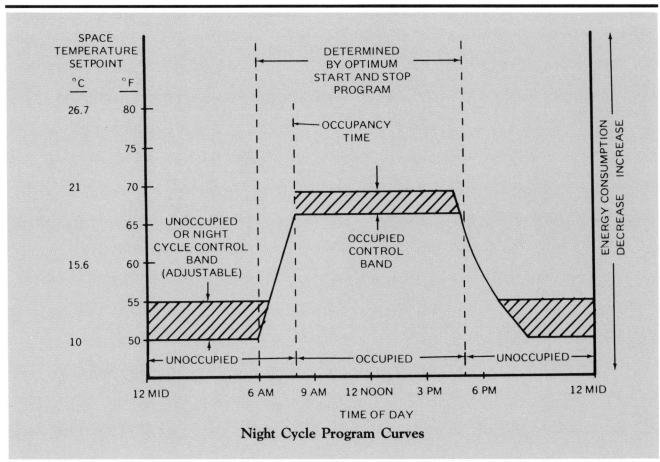

Figure 2.25

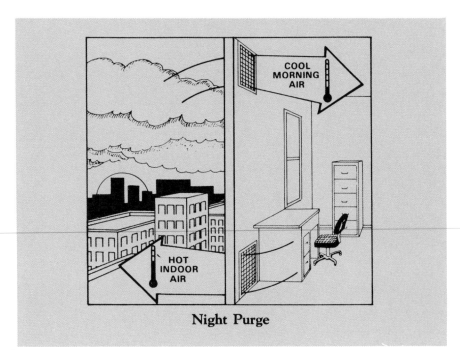

Figure 2.26

In reheat systems, space sensors in every zone monitor zone temperatures. This information is fed to the program, and the HVAC unit discharge air temperature is reset accordingly. This same reset strategy is applied to the hot and cold deck of a dual duct system and multizone units. The program may also be applied to reset chiller discharge water temperature as a function of outdoor air, or zone requirements, or to a hot water boiler, or heat exchanger. Figure 2.28 gives an overview of the load reset program.

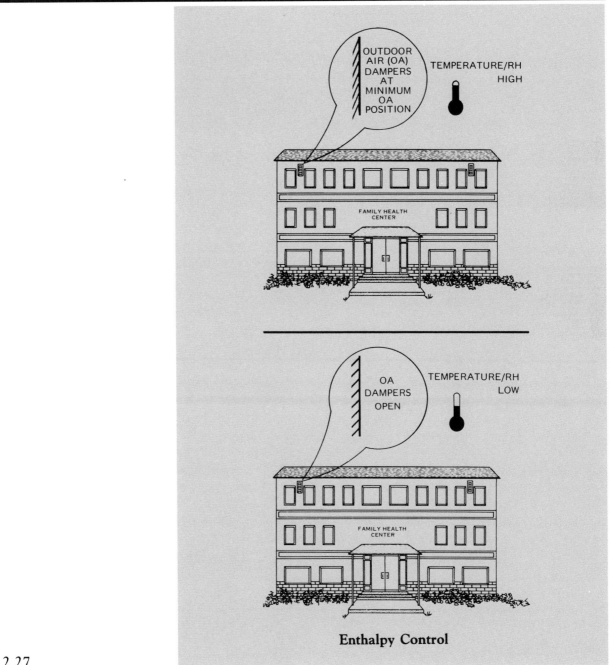

Enthalpy Control

Figure 2.27

Zero-Energy Band Program

This program applies to HVAC systems that have heating and cooling capabilities. It eliminates energy waste resulting from simultaneous heating and cooling of air delivered to the space. The space comfort range is divided into three categories: heat, zero-energy band, and cool. These categories are adjustable and may be set as desired by the building occupant. At the lower end of the comfort range is the heating region. When the space temperature enters this region, the heating source is energized. At the upper end of the comfort range is the cooling region. When space temperature is in this region, the cooling source is energized. Between the heating and cooling regions is the zero-energy band, within which region both heating and cooling sources are off.

The zero-energy band program interacts with the load reset program. When the space temperature is outside of the heating and cooling regions, the heating and cooling sources are de-energized. The program, however, continues to monitor space temperatures and perform load reset. The heating and cooling remain de-energized until the space temperature re-enters one of these regions.

When an HVAC system, such as a multizone unit, serves different exposures (north and south), or different internal loads (interior and exterior), temperatures from zone to zone may vary considerably. The

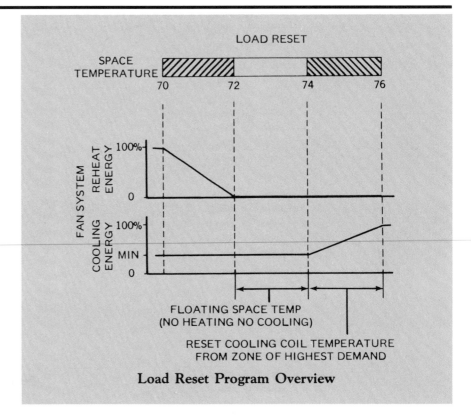

Load Reset Program Overview

Figure 2.28

program constantly scans zone space temperatures and makes cooling decisions based on the highest temperature, and heating decisions based on the lowest temperature. This ensures that all zones will remain in the comfort range (See Figure 2.29).

Energy Management – Lighting

Lighting control falls into two basic categories: (1) simple on/off for occupied/unoccupied control, and (2) actual control of lighting levels in a building during occupied periods.

Centralized control of lighting is most conveniently accomplished when it is designed into new building electrical systems. Contactors can be installed ahead of lighting circuit breaker panels for each floor, zone, or area of a building. Emergency lights, wall outlets, and non-lighting circuits can be tied into circuit breaker panels not under control of the lighting program.

Adapting lighting control to existing buildings is not always economically feasible. Frequently, the existing circuit breaker panels carry a mixture of loads, such as lighting, wall outlets, exhaust fans, etc. This would require removing the lighting circuits and wiring them through a contactor to a new circuit breaker panel. Another option would be to install a low voltage lighting control system. The final decision would have to be made on the basis of economy and return on investment.

Occupied – Unoccupied Program

Programmed lighting applies the basic on/off, or occupied/unoccupied programming strategy to building lighting systems. It ensures the owner that lights automatically turn off at the end of normal occupied periods. Program override capability is available for unusual or

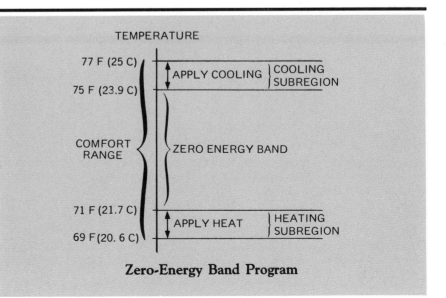

Zero-Energy Band Program

Figure 2.29

overtime periods of occupancy. This override can be applied only to the zones to be occupied, thereby leaving all other zones or floors in the unoccupied mode.

Lighting Level Control

Lighting level control complements the occupied/unoccupied program described above. It controls the level of lighting during occupied periods. There are two methods of controlling lighting levels. The first method uses multiple lamp fixtures to provide different light levels. Two- or four-lamp fixtures are divided into two circuits to provide either one half capacity or full-on lighting levels. Three lamp fixtures can provide one third, two thirds, or full-on lighting levels.

The second method applies primarily to fluorescent lighting and is most effective in open landscape lighting design. Ambient lighting through windows or skylights is also a requirement to supplement the artificial lighting. Under this program, an ambient light sensor measures the total light in the space (daylight and artificial light), and increases or decreases artificial light levels to maintain total light levels according to predetermined standards (See Figure 2.30). These lighting control programs can reduce overlighting, reduce energy consumption and demand charges, and reduce air-conditioning costs.

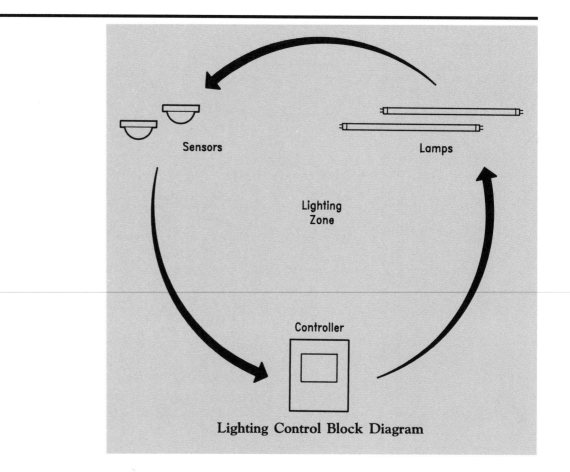

Figure 2.30

Lighting Control Block Diagram

Fire Safety Systems

Stand-alone fire alarm systems can be integrated into a distributed process system, thereby providing enhanced alarm capabilities and fire safety programs. This is particularly advantageous to the existing building which may have an existing, functioning, in-house fire alarm system. In a new building, the owner or engineer may want to totally integrate the fire safety system into the Building Automation System, or may prefer to install a stand-alone fire alarm system with redundant alarming into the BAS. In both cases, the stand-alone system and the totally integrated system, the benefits are substantial. Life safety and asset protection are probably two of the major benefits to be derived from either option. The subject of fire safety systems will not be covered in detail in this book; however, a general overview of capabilities will be offered. The reader should consult current NFPA Standards, local building codes, and local authorities having jurisdiction over fire safety systems before proceeding with system design.

Detection Methods

Alarm initiating devices for fire alarm systems can be manual pull stations, heat detectors, or sprinkler system waterflow switches. Smoke detection systems may consist of area or duct smoke detectors. These may be either ionization or photoelectric type units. Flame detectors in the form of infrared sensors are also available. These sensors respond to infrared radiation which is emitted from a flickering flame. National Fire Protection Association standard 72E contains guidelines for the placement of detection system elements.

Annunciation and Alarm

Upon actuation of any alarm initiating device, any or all of the following annunciation and/or alarm conditions may occur:

- An English language announcement of the alarm and an indication of its zone or detector location will appear on the host computer CRT. An audible alarm will sound, and the alarm condition will flash on the CRT to notify the operator of the emergency. Emergency action instructions will be displayed on the CRT screen for the operator's information.
- The host computer printer will document the alarm by type, zone, or actual detector location, time and date.
- The municipal alarm system will be actuated.
- Audible and visual alarm devices will be activated throughout the building. They will remain active until the alarm initiating device is reset and the system is returned to normal through the host computer terminal.
- Prerecorded voice action messages may be broadcast to building occupants over speakers located throughout the building. These may be specific instructions directing building occupants to safe locations.
- Preprogrammed smoke control procedures can be initiated. Such procedures could involve the activation of HVAC fans, exhaust fans, and dampers to clear smoke from the fire location and pressurize adjacent areas to prevent smoke migration. Self-closing smoke and fire doors may also be released.
- The elevator call system may be disabled and all elevators returned to the ground floor for manual operation by firefighters. If the fire is in a ground floor zone, the elevators can be returned to another floor.

- Prerecorded voice messages can be transmitted to elevators.
- Doors which are locked by the access control (security) system may be programmed to unlock to provide occupant egress. This function may also be performed by the operator from the host computer terminal.

Building Security/Access Control

Access control systems can be integrated into the distributed process system along with HVAC and fire alarm systems. These access control systems control personnel access to secured areas or the entire building. Credit card-sized access cards with a coded magnetic stripe or barium ferrite layer are assigned to authorized personnel. Cards may allow the cardholder access to a specific room, through specific doors, and within certain established time limits. If the access control system is applied to elevator control, the cardholder's access may be limited to specific floors.

When the card is inserted, or swiped (passed through a card reader), the system reads the ID number on the card in conjunction with the location, the time, and the day. This data is compared with the preprogrammed, valid information stored in memory. If the comparison is positive, indicating an authorized entry or exit, a signal is sent to activate the appropriate door strike. The host computer printer will document the card ID number, card reader location, and time of day. If there is not a match with the preprogrammed conditions, the door will remain locked and a printout will occur indicating the reason why access or egress was denied. For additional security, an electronic keypad may be installed to act in concert with the card reader. In this case, the cardholder must enter a four-digit ID number and insert the card to obtain authorized entry. This prevents unauthorized use of stolen or lost cards.

In most systems, if the host computer fails or the communications bus to the access control panel fails, the access control panel will maintain control. It will also store all card activity, alarms, time, and day information in memory. When communication with the host computer is restored, all stored information will be uploaded for documentation purposes.

If any door designated for card reader access is operated without the use of a valid card, an alarm condition will display on the host computer CRT and printer for operator action.

Central Host Computer

The complete distributed process system consists of the following:
- A central host PC as the operator interface
- Distributed processors (controllers)
- Remote sensors, detectors and actuators

The distributed processors (controllers) connect directly to the remote sensors, actuators, and detectors to provide constant control functions, energy management strategies, and fire/security monitoring. The processors contain software programs that accept information from these remote devices, process the information, and perform control functions. The processors communicate with one another over a communications bus by means of a common language, or protocol. Although these processors are intelligent, and can perform their functions on a stand-alone basis, the central personal computer can provide high level system monitoring, control, alarming, and management reporting. In summary, the central PC provides user-friendly operator interface capabilities.

Building Management Programs

Most distributed process systems have standard application packages and optional software programs which enhance the capability of the personal computer as a building operations tool. These application packages and programs provide reports and documentation to assist building owners and operation personnel in managing building systems in the safest and most cost effective manner.

Historical Data

The building operator may initiate a program which will store or retrieve information relating to a specific point, or group of points. He/she can manually request a printout of the information, or program an automatic printout of the reports on a regular, periodic basis. The reports may include alarms, return-to-normal, points in trouble, or point status (temperature, humidity, on/off, etc.). These reports allow the operator to recognize trends and, in many cases, predict a malfunction before it occurs. A sample of a historical trend log printout is shown in Figure 2.31.

HISTORICAL TREND LOG REPORT

4TH FLOOR ADMINISTRATION AH-6 PERFORMANCE

Starting Date and Time: Wednesday Mar 13,1985   22:30
Ending Date and Time : Thursday Mar 14,1985   00:10
Trend log Interval   : 00:10

01 AH-6 SYSTEM TEMPERATURES
02 AH-6 SYSTEM TEMPERATURES
03 AH-6 SYSTEM TEMPERATURES
04 AH-6 SYSTEM TEMPERATURES
05 AH-6 SYSTEM TEMPERATURES
06 AH-6 FAN STATUS
07 AH-6 FAN STATUS

T/D	Item 01	Item 02	Item 03	Item 04	Item 05
22:30	060.5	062.6	062.0	061.8	064.3
03/13/85	Deg	Deg	Deg	Deg	Deg
22:40	060.1	062.3	062.2	061.8	064.3
03/13/85	Deg	Deg	Deg	Deg	Deg
22:50	059.6	062.0	061.9	061.5	064.1
03/13/85	Deg	Deg	Deg	Deg	Deg
23:00	059.0	062.1	061.9	061.2	064.0
03/13/85	Deg	Deg	Deg	Deg	Deg
23:10	059.0	061.8	061.7	061.0	063.8
03/13/85	Deg	Deg	Deg	Deg	Deg
23:20	058.7	061.6	061.4	060.7	063.9
03/13/85	Deg	Deg	Deg	Deg	Deg
23:30	058.5	061.3	061.1	060.2	063.5
03/13/85	Deg	Deg	Deg	Deg	Deg
23:40	058.2	061.3	060.7	060.3	063.3
03/13/85	Deg	Deg	Deg	Deg	Deg
23:50	058.3	061.2	060.7	060.2	063.0
03/13/85	Deg	Deg	Deg	Deg	Deg

Historical Trend Log Print-out

Figure 2.31

Action Messages

Preprogrammed messages that print and display on the CRT are an important feature when a point goes into alarm or trouble. Regardless of the operator's skill level, this feature provides an element of support in emergency situations. This is extremely important for fire and security applications.

Energy Audit Program

Under this program, a building owner may use a spreadsheet format to convert specified measured data (kilowatt hours, cubic feet of gas, gallons of oil, etc.) into logical management reports. The output can be a simple listing of energy consumption data for a given period, or the raw values may be converted to management-related terms, such as kilowatt hours per square foot, or cubic feet of gas per degree day. This allows building owners or managers to better understand the energy requirements of their building and the effectiveness of their energy management strategies (See Figure 2.32).

Material Inventory

A material inventory program provides a convenient method of inventory control for the building maintenance or facilities department. This program can track material quantity, cost, usage, and if desired, it

```
PARKWAY BUILDING                        BUILDING ENERGY AUDIT
320,000 SQ FT                           REPORT FOR MARCH 1991

                              THIS MONTH

                        1990        1991      % CHANGE
ELECTRICITY
     KWH              790,435     649,330       -18
     KWH COST ($)      45,924      44,090        -4
     PEAK DEMAND (KW)   3,525       3,201        -9
     DEMAND COST ($)   14,276      16,645        17

NO. 2 OIL
     GALLONS           10,507       9,141       -13
     TOTAL COST ($)    11,663       8,958       -23
     MILLION BTU        1,471       1,280       -13

TOTAL ENERGY
     COST ($)          57,587      53,048        -8
     MILLION BTU        4,168       3,495       -16

PERFORMANCE
     HTG DEGREE DAYS      891         849        -5
     CLG DEGREE DAYS        0           0         0
     BTU/SQ FT         12,944      10,854       -16
```

Energy Auditor Print-Out

Figure 2.32

can monitor material location. Inventory minimum levels can be preprogrammed into the computer. When levels drop below these levels, a reorder printout will be generated. This ensures that materials and parts are always available when needed (See Figure 2.33).

Maintenance Management Program

The maintenance management program is a valuable tool for facilities management personnel. This program documents the required maintenance tasks, the level of skill required for each, and the tools and materials necessary to perform each task. Scheduling of preventive maintenance activities can be based on either calendar time or equipment run time or event occurence. Preventive maintenance printouts may be generated on a daily basis. These printouts or work orders include the following information:

- Purpose of maintenance task and instructions
- Description of piece of equipment
- Type of worker required (electrician, mechanic, etc.)
- Normal time requirement to perform task
- Materials and tools required to perform task

1991/05/20

MAINTENANCE MANAGER
****

MATERIALS TO REORDER
****

MATERIAL ID	UNIT	QUANTITY ON HAND	COST ON HAND	QUANTITY TO REORDER	ESTIMATED PRICE
MAT01	EA	100.00	$ 200.00	200.00	$ 2.00
MAT04	EA	2.00	$ 250.00	10.00	$150.00
MAT05	EA	100.00	$ 200.00	200.00	$ 2.00
MAT06		5.00	$ 170.00	30.00	$ 34.00
MAT08	DOZ	12.00	$4,200.00	20.00	$350.00

*** END OF REPORT ***

1991/05/20

MAINTENANCE MANAGER
****

MATERIALS ON ORDER
****

MATERIAL ID	UNIT	QUANTITY ON HAND	COST ON HAND	QUANTITY TO REORDER	ESTIMATED PRICE
MAT02	EA	100.00	$ 200.00	200.00	$ 2.00
MAT07	EA	5,000.00	$50,000.00	50.00	$10.00
MAT10	GAL	13.50	$ 1,188.00	25.00	$88.50

*** END OF REPORT ***

Inventory Control Report

Figure 2.33

After completion of the task, the actual time and materials used are entered into the system. The program will track ongoing costs for each item (task, materials, skill) on a monthly, yearly, or lifetime total basis. A printout of incomplete tasks can be generated to check work in progress. This information assists supervisors in determining workload and scheduling personnel.

To protect sensitive data that may be stored in the maintenance management program, operator access can be limited to different levels. For instance, a lower level operator may have access to work orders and management reports only, whereas a higher level operator can not only access this data, but also modify maintenance information in memory (See Figure 2.34).

Color Graphics and Annunciation

Color graphic software presents graphic displays of HVAC system schematics, building floor layouts, bar charts, or graphs. This is a convenient way of presenting pertinent information in a multi-colored, logical format for the system operator. The multiple colors can give greater meaning to graphic displays and emphasize critical data. For example, normal point status might be displayed in green, and alarm status displayed in red. The operator can call up a system graphic and review large amounts of data relating to that graphic. The data is dynamic in nature, which means that it changes as actual changes occur.

With most systems, the operator can create custom graphic displays while the system is operational. The operator selects standard symbols and HVAC system diagrams from a library of multicolored graphics. Standard libraries contain HVAC equipment symbols for items such as fans, pumps, dampers, valves, sensors, etc. Fire and security symbols display fire and smoke detectors, manual stations, bells and horns, card readers, etc. Floor plans, piping schematics, and ductwork layouts are also available.

Selected analog data can be programmed to create graphs or bar charts. This is beneficial to building owners and operations personnel in comparing past and present performance. Dynamic graphs and bar charts for selected points can be displayed on an ongoing basis. This provides the operator with up-to-the-minute system information and trends so that corrective action can be taken if required.

Voice Communications

Most manufacturers now offer off-site voice communications capability. This means that a building owner or authorized person can dial in to the host computer from a remote touchtone telephone, and request information or change point settings. The caller must have a valid password to enter the system. The password will be requested verbally when the proper telephone number is dialed. The caller then enters the password through the touchtone pad. The caller may then request the status of a specific point (through a code entered through the touchtone pad) and the requested information will be returned verbally, along with the point action message. If the caller wishes to make a set point change, he/she can do so on the same call. Access to specific points can be restricted to specific password holders.

```
 WORK ORDER NUMBER 87 - 13 - B

    PRINTED BY HONEYWELL, INC. MAINTENANCE MANAGER on 08/11/87

                              FOR

                        "SAMPLE SYSTEM"


 EMERGENCY WORK ORDER     (CT SMA # 00001)

 Complete by:
     08/12/87

 Work Order Purpose:
     fan system periodic maintenance

 To be performed on:
     23.01     SUPPLY FAN STATUS

 Performer:
     mechanic

 Task Purpose:
     replace belt

 Instructions:
     turn off power- remove belt
     install belt- turn on power

 Estimated completion time:
     45  minutes

 Materials Required:
     1   unit(s) of fan belt

 Tools Required:
     wrench

 ===========================================================================

 COMPLETION INFORMATION

 Yes   No

 ---   ---   Took Estimated minutes to complete.
             If no, how many?  _____

 ---   ---   Used Estimated material quantity.
             If no, how much?  _____

 COMMENTS:


 _____    _____   _____
 Name                                Date             Time
```

Maintenance Task Print-Out

Figure 2.34

Dial In/Out Function

In most systems, two-way communications are possible between the host computer and the DDC controllers over voice grade telephone lines. Through the use of modems, controllers located in remote buildings can direct dial the host computer to report off-normal alarms. The host computer can periodically dial the remote controllers and request the transmission of all point data and alarms.

Building owners or facilities personnel who have a compatible PC and software, at a remote location can monitor and operate the Building Automation System at their building. They can verify HVAC system operation and make adjustments to set points from the remote location. This off-site monitoring and control is an attractive feature to many who are responsible for building operations.

Chapter Three

System Architecture

Building Automation Systems centralize the operation, monitoring, and management of building systems, such as heating, ventilating, air-conditioning (HVAC), lighting, fire safety, and security-access. The goal of these systems is to provide a comfortable and healthy environment for building occupants, reduce energy costs, and maintain a high level of life safety and security. These systems enable building owners and operations personnel to operate their facilities more efficiently, while making the most effective use of maintenance and operations staff.

The first Building Automation Systems (BAS) were basically electro-mechanical, whereby sensors and controlled devices were hard-wired or piped to a large central control panel. Over the past 25 years, however, these systems have become computerized. Today, a BAS consists of *remote sensors and actuators, microprocessor-based controllers*, an optional *central host computer (PC)*, and an *optional management host computer*. The remote sensors (Level I) sample the building environment and transmit pertinent data to the controllers. The controllers (Level II) perform Direct Digital Control (DDC) functions and energy management strategies by controlling remote actuation devices. Building management reporting, monitoring, and access to system data can be accomplished through the controllers, the central host PC (Level III), or an optional management host computer (Level IV). These BAS components are illustrated in Figure 3.1 and described in detail in the following sections.

Remote Sensors and Actuators (Level I)

The sensors, actuators, and controlled devices that are connected to Level II controllers are known as points. Points are classified by their input or output function as follows:

- Digital inputs (DI)
- Digital outputs (DO)
- Analog inputs (AI)
- Analog outputs (AO)

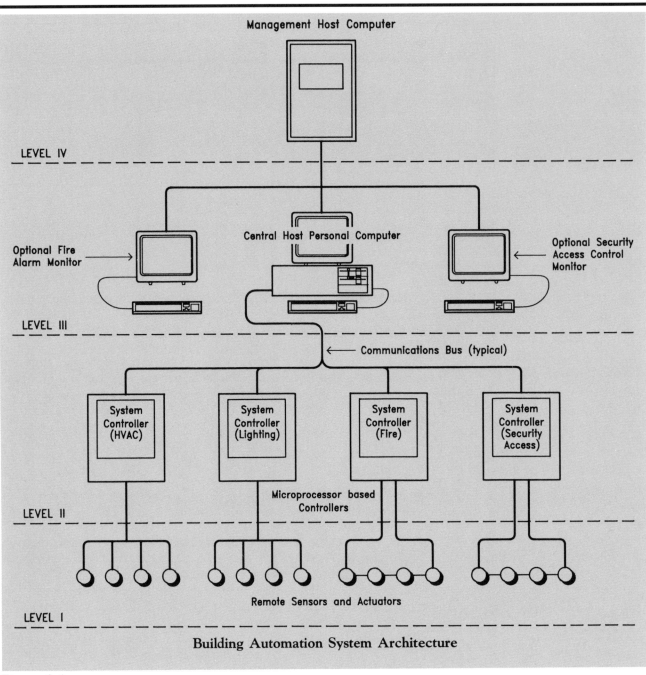

Figure 3.1

Digital Inputs (DI)

A *digital input* point is an external sensor that sends a two-state, on-off, or two-position signal to a controller. An example of this would be electrical switch contacts, which are either open or closed. Low-temperature detection controllers, high-temperature detection controllers, duct smoke detectors, or duct static pressure sensors are all considered digital input points. The duct static pressure sensor shown in Figure 3.2 has built-in switch contacts that close when the duct static pressure exceeds its setting, and open when the duct static pressure drops below its setting. This device is a digital input sensor used to signal the controller that a fan is or is not operating.

Digital Outputs (DO)

A *digital output* point is the two-position switching action of switch contacts residing in the controller. This switching action makes or breaks an external electrical circuit. The circuit could be wired to turn lights on or off, or to start and stop an electrical motor such as a supply fan for an air handling unit shown in Figure 3.3.

Analog Inputs (AI)

An *analog input* point is an external device that sends a proportional or variable signal to the controller. Examples may be a humidity sensor, water or duct airflow sensors, or a space temperature sensor, as shown in Figure 3.4. Unlike digital input points, analog input signals are sent over the entire range of the sensing device. Most controllers will accept analog input signals of 2 to 10 volts DC (VDC), 4 to 20 milliAmp (MA), or a variable resistance.

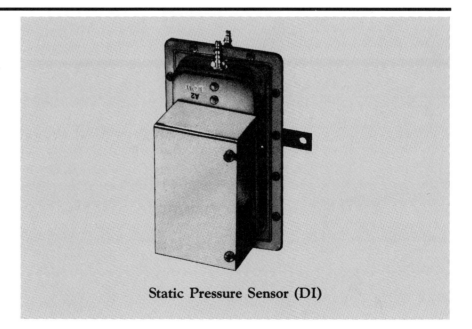

Static Pressure Sensor (DI)

Figure 3.2

Analog Outputs (AO)

An *analog output* point is a proportional, variable signal sent by the controller to modulate an external control device. The analog output signal can be 2 to 10 VDC, 4 to 20 MA, or 3 to 15 pounds per square inch gauge *(PSIG)* air pressure. The external controlled device could be a variable frequency drive, a damper actuator, or a valve actuator, as shown in Figure 3.5.

A typical example of a basic five point control system, using the four types of points employed by direct digital control systems, is shown in Figure 3.6.

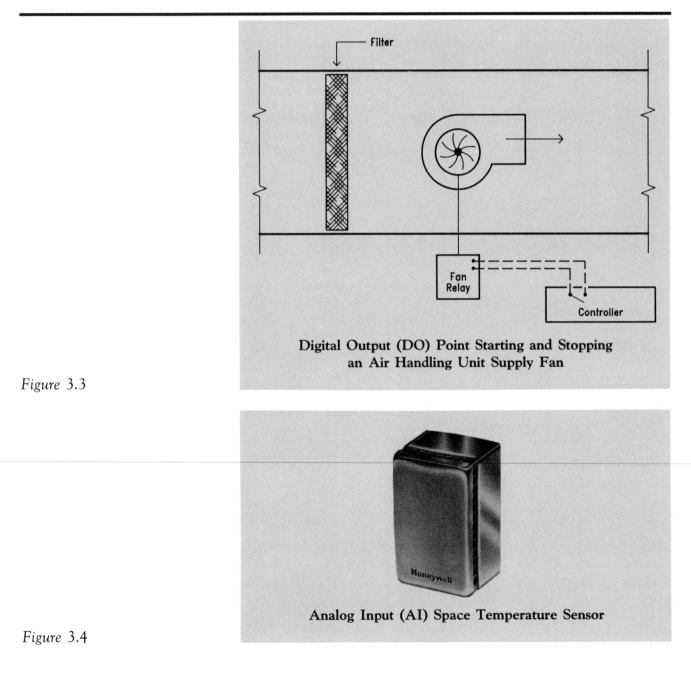

Digital Output (DO) Point Starting and Stopping an Air Handling Unit Supply Fan

Figure 3.3

Analog Input (AI) Space Temperature Sensor

Figure 3.4

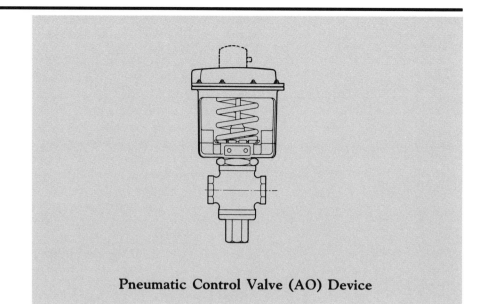

Pneumatic Control Valve (AO) Device

Figure 3.5

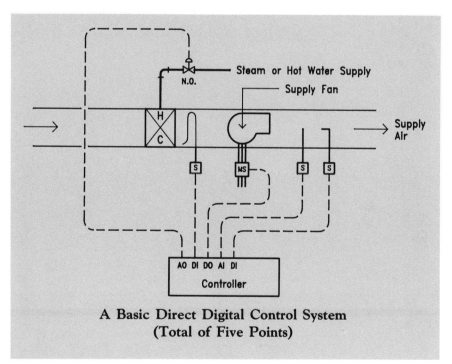

**A Basic Direct Digital Control System
(Total of Five Points)**

Figure 3.6

The sequence of operation for this system would be as follows:

1. The controller energizes the supply fan magnetic starter (MS) by a digital output point.
2. A digital input point – a static pressure sensor located in the supply air duct – closes its switching contacts in response to an increase in static pressure, indicating to the controller that the fan is operating.
3. An analog input point – a temperature sensor located in the supply air duct – sends a variable input signal to the controller.
4. The controller maintains its setting by sending an analog output signal to modulate the automatic control valve.
5. If the heating coil discharge air temperature drops below the low temperature detection sensor setting, the sensor opens its contacts, sending a digital input signal to the controller. The controller sends a digital output signal to stop the fan, and an analog output signal to open the automatic control valve.

Microprocessor-Based Controllers (Level II)

In general terms, a controller receives a signal from a sensor (Level I), compares the signal with a preset set point, and determines if corrective action is necessary. Corrective action is in the form of an output signal to modulate or two-position an actuator or controlled device (Level I).

Controllers are microprocessor-based. Since their software programs are in digital form, the controllers perform direct digital control. Microprocessor-based controllers can be used as stand-alone controllers or can be incorporated into a Building Automation System.

Microprocessor-based controllers offer a more flexible and accurate means of control than traditional electric, electronic, and pneumatic control systems. Through the use of software programs, the controllers can provide a variety of HVAC control functions, energy management programs, and other building management functions. Revisions to control sequences can be readily accomplished without major hardware changes, simply by changing the program software.

Since the controllers are microelectronic, they contain no moving parts and, therefore, their maintenance requirements are minimal. The controllers are reliable, and most manufacturers include self-diagnostic features to notify maintenance personnel of any malfunctions.

In addition, most Level II controllers are provided with a communication port or channel for connection to a remote or portable operator's terminal, printer, or telephone modem. Figure 3.7 illustrates a printer connected to a Level II system controller.

In general, most manufacturers offer two basic types of controllers: *zone controllers* and *system controllers*. These are illustrated in Figure 3.8 and described in the next sections.

Zone Controllers

Zone controllers typically control HVAC terminal units, such as variable air volume (VAV) terminal units, fan-coil units, heat pumps, smaller air handlers, packaged heating/cooling units, and laboratory fume hoods. Usually these controllers have relatively few connected Level I points and standardized control sequences, and are designed for specific applications. For example, a VAV terminal unit zone controller is frequently packaged with an internal damper actuator (Level I), and has only the required point capacity to meet the specific application. The DDC software program is resident in the controller.

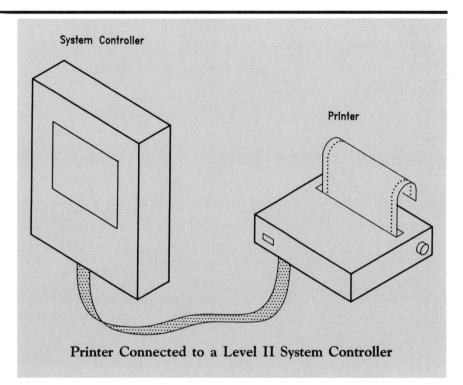

Printer Connected to a Level II System Controller

Figure 3.7

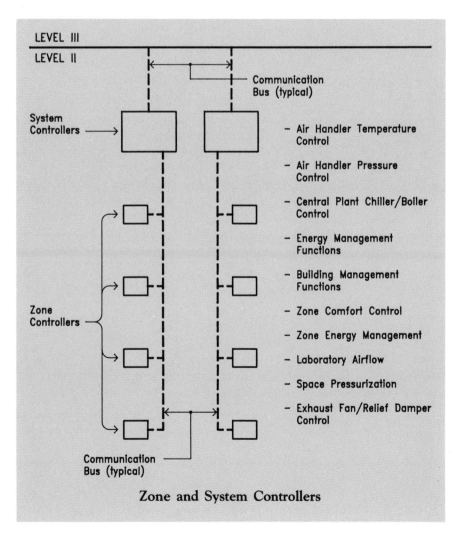

LEVEL III

LEVEL II

Communication
Bus (typical)

System
Controllers →

— Air Handler Temperature
  Control

— Air Handler Pressure
  Control

— Central Plant Chiller/Boiler
  Control

— Energy Management
  Functions

— Building Management
  Functions

— Zone Comfort Control

— Zone Energy Management

Zone
Controllers

— Laboratory Airflow

— Space Pressurization

— Exhaust Fan/Relief Damper
  Control

Communication
Bus (typical)

Zone and System Controllers

Figure 3.8

In general, the controller has a communication port, or channel, connected to a portable operator terminal for use during initial setup and for subsequent adjustments. A communications bus enables the networking of zone controllers, allowing Level I point information to be shared with other system controllers and with a central host PC (Level III). (Communication buses are discussed in more detail later.) For example, a system level controller can use space temperature values from a zone controller as inputs to perform optimum start, optimum stop, night setback, and night purge programs. A Level III central host PC can use the same Level I point information, in text or graphic display form, as data for trend reports, alarms, and historical logs and files.

Typical Point Data

For a typical VAV terminal unit controller, the following Level I point and DDC program information is generally available for display or program use:

- Airflow in cubic feet per minute (CFM)
- Set point temperature
- Deviation from set point (offset)
- Space temperature
- Duct air temperature
- Damper position
- Reheat stages energized (for electric and/or fan reheat coils)
- Hot water valve position (for hot water reheat coils)
- BTU per hour added to or removed from the space
- Maximum/minimum CFM set point

Other types of zone controllers provide different information, depending on the application.

System Controllers

System controllers have a greater capacity than zone controllers in terms of the number of Level I points, DDC loops, and control programs. System controllers are usually applied to major mechanical HVAC equipment, such as large built-up air handlers, central VAV systems, and central heating and chiller plants. Other types of system controllers are designed to monitor individual or multiple zones of fire alarms, unauthorized intrusion alarms, or to provide access or lighting control. They can provide smoke containment or evacuation programs, emergency occupant evacuation control, or can control personnel movement through card reader access.

Controllers at this level interface with controlled equipment (Level I) directly through Level I sensors and actuators, or indirectly through communications links with other system or zone controllers. System controllers typically have a communications port or channel to accept a portable operator's terminal during initial setup, and for subsequent adjustments. When the system controllers are linked to a Level III central host PC, changes to the controller software program can be made as follows:

- Software program changes are performed at the Level III central host PC and down-line loaded to the system controllers over the communications bus.
- Software program changes are performed at the system controller and are up-loaded to a Level III central host PC.

Level I point information, processed by system controllers, is used directly by the resident DDC, EMS (Energy Management System), and Time/Event software programs. The information is accessible, either through a keyboard access and display permanently mounted on the system controller door, or a portable operator's terminal. This information can be communicated to other system controllers and to a Level III central host PC. Programmed parameters (such as set points, commands, and overrides) and Level I point values associated with DDC and other resident programs can also be read and adjusted through a local keyboard and display, a portable operator's terminal, or a Level III central host PC. However, system controllers will continue to operate as programmed on a stand-alone basis should the communications link with other system controllers or the Level III central host PC be lost.

Communications Bus

The communications bus links controllers to one another as well as to a Level III central host PC. The most common communications bus media are twisted shielded copper (called *hard-wire*), *direct-dial telephone lines*, and *fiber optic cable*. Some factors to consider when selecting a medium are the signal to be transmitted, cost, geographic layout, and the possibility of electrical interference.

Hard-Wire

Twisted shielded copper conductors (hard-wire), ranging in size from 16 to 24 gauge, is the most commonly used and the most economical choice for single building applications. Communications bus lengths of up to 4,000 feet are common, without the use of auxiliary extenders or repeaters. When repeaters are used, extension of three to four times this distance are possible. The two most common communications bus wiring configurations are daisy chain (Figure 3.9) and star (Figure 3.10).

Direct-Dial Telephone Lines

Common carrier direct-dial telephone lines can link distant buildings. Dial-up to a single remote Level II system controller is possible with the proper telephone modems, as shown in Figure 3.11.

Fiber Optic Cable

A fiber optic communications bus is particularly well-suited to installations in an environment that interferes with its communications, such as areas subject to radio frequency interference (RFI) generated by electrical power lines or electrical storms. The disadvantages of fiber optic transmission are high material and installation costs and the absence of established guidelines for this technology.

Central Host Personal Computer (Level III)

The central cost personal computer acts as an operator-machine interface. It provides an operator with a way to access program information, point information, and management reports, and to change programs when necessary. It also allows the operator to monitor equipment status, receive alarms, and take corrective action when alarms occur.

The PC consists of an operator terminal with color CRT, self-prompting system access, memory expansion capability, and communication boards. The PC also contains plug-in function boards and multiple communication ports or channels to accommodate peripheral devices such as a telephone modem and local or remote terminals and printers.

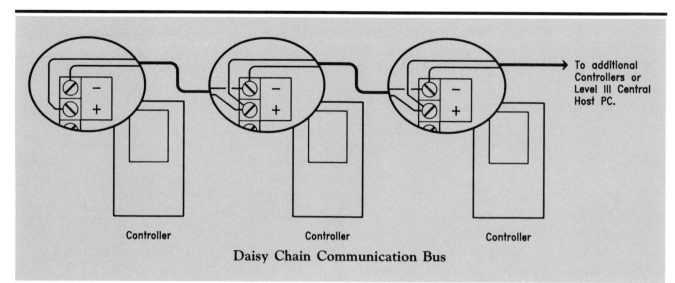

Daisy Chain Communication Bus

Figure 3.9

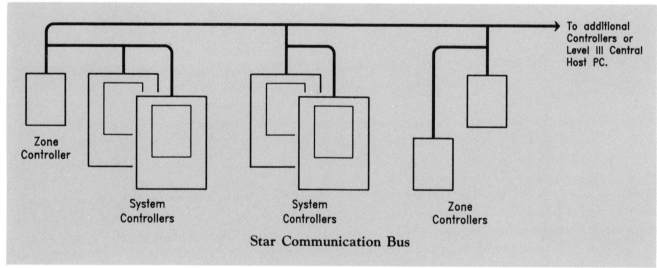

Star Communication Bus

Figure 3.10

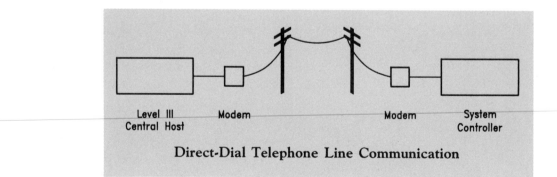

Direct-Dial Telephone Line Communication

Figure 3.11

Some application software features are as follows:

- *Multiple level access codes*: Operators are authorized access to specific levels of programming. This prevents unauthorized tampering with software or system operation. It also ensures that only authorized personnel will have access to proprietary information.
- *Customized programming capability*: Custom programs can be written for specific sequences and applications. The programs can be entered through the PC operator's terminal and down-line loaded to specific system controllers.
- *Custom colorgraphic displays*: The CRT can display multicolored, dynamic layouts of HVAC systems, piping diagrams, lighting systems, fire/smoke detectors, and floor layouts. It can also display dynamic bar charts and graphs to provide building management personnel with an ongoing picture of building costs, energy consumption, or mechanical equipment efficiencies.
- *Automatic or manually requested log report printouts*: The system can be programmed to periodically print logs of:
  - All points status
  - All points in alarm
  - Single point status – to observe trends
  - Unacknowledged alarms
  - Energy consumption
  - Building access time and access card number
  - Historical trends
- *Maintenance Management Program*: Preventive maintenance can be scheduled, based on run time of equipment, calendar time, or an event occurrence. Work orders will show estimated task time, tools required, and classification of worker.
- *Customized Reporting Program*: Point and controller information can be assigned to report and display, along with action messages for the operator, at specific CRTs and printers. Security and fire points can be designated for one CRT and printer, and HVAC-related points to another CRT and printer.
- *Software integration of building systems*: Interactive software programs can integrate HVAC, fire alarm, security, access control, and lighting systems.

Management Host Computer (Level IV)

Level III central host PCs are the most commonly used as an operator-machine interface to a BAS. However, Level IV management host computers may be used for applications, such as campuses with multiple buildings or complexes with multiple remote facilities. Each of the buildings may have its own independent central host PC.

Another example is a single building organization that is using its PC extensively. If facility management wishes to change programs or request management reports, this may be extremely difficult when the system is so actively used.

In the case of multiple buildings with separate PCs, there may be a need by facility management for a centrally located PC to receive management reports and/or critical alarms. The Level IV management host computer fills these needs.

Management host computers are at the top of the BAS hierarchy. They are used to manage and execute commands to lower level devices. In general, their primary role is to retrieve, archive, and process building point information, such as energy consumption, operating costs, and alarm activity. This information allows the computer to generate customized reports, such as graphs, bar charts, and spreadsheets necessary for the long-term management and operation of the building. At this level, facility management personnel can access any or, in some cases, all system point information, and execute commands to operate the system.

A powerful personal computer (PC) can be adequate for the management host computer function. Depending on the requirements for memory size, networking, and data processing (such as processing speed), software compatibility and capability, a mini-computer might be used, with hard disk storage. Peripheral devices, such as terminals, keyboards, CRTs, and printers are the same as those listed for a Level III central host. The software used for management Level IV and central host Level III computers is also identical, with the exception that the management level reports and graphic capability needs of Level IV are more comprehensive than those required for Level III. Many management level applications can be integrated into the Level III central host computer. However, in the case of BAS applications in extensive and/or diverse facilities or campuses, certain management functions may necessitate a Level IV computer. Some of these management functions and duties may include:

- The up or down transfer (within the BAS hierarchy) of operator authority and assignments during regular or unattended periods of building operations.
- Logging and reporting of any changes to BAS, DDC, EMS, or any other software programs identified by operators.
- Reports and summaries, over any user-defined time period, of all acknowledged and unacknowledged alarms; and facility management information, such as energy consumption, heating and/or cooling plant efficiency.
- Historical reports of all alarm occurrences and their return to normal.
- Archiving of all system controller DDC, EMS, and other software programs on back-up storage media.
- Facility cost allocations and accounting of energy consumption to specific buildings, systems, departments, or tenants.
- Daily reports of all BAS hardware, firmware, software, and equipment failures and their return to normal.
- Operator and maintenance means to oversee, and fine tune a facility's most critical systems on a dynamic basis.
- Segregated alarms pertaining to a facility's most critical systems.

Integrating Other Building Systems

A building owner or manager may wish to integrate other building systems, such as fire alarm, security, access control, and lighting control, with his HVAC systems. There are two methods of integrating these systems into the Building Automation architecture. Fire alarm, security, and access control systems are basically stand-alone systems. They each have the equivalent of a controller, Level I sensors, and actuators. These systems have central alarm panels which are connected to remote sensors, detectors, and alarm devices. The central panels are considered controllers (Level II), while the sensors, detectors, and alarms are Level I devices. Figure 3.12 shows how a stand-alone fire alarm system can be integrated with an HVAC system, with all point information displayed and accessible at the central host personal computer (Level III).

In this example, the fire alarm panel contains auxiliary alarm contacts and status contacts (digital outputs) which are hard-wired to a standard system controller as digital inputs. In the event of an alarm, the auxiliary contacts close, signalling the alarm condition to the system controller. The system controller can initiate preprogrammed smoke containment or evacuation programs, or building occupant evacuation procedures. The fire alarm panel displays the alarm, and if it is connected to a proprietary central fire alarm computer, the alarm will be displayed there also. The system controller also communicates the alarm to the BAS central host computer. The central host printer generates a printout of the alarm information and the CRT presents a display of alarm information for operator action.

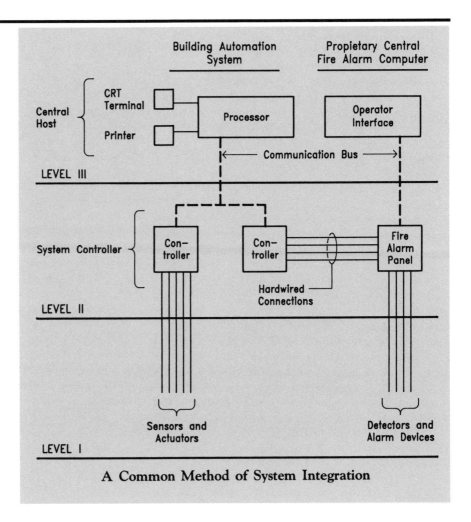

A Common Method of System Integration

Figure 3.12

This method of integrating HVAC and other building systems applies primarily to existing buildings. Many of these buildings have up-to-date, functioning fire, security, or access control systems. If a BAS is to be installed, it may be advantageous, from a building management standpoint, to integrate these systems with the HVAC systems.

Figure 3.13 illustrates another method of integrating building systems. HVAC zone and system controllers are linked to the central host computer by a communications bus. Another communications bus connects the fire or security controller to the host computer. These controllers are designed specifically as fire alarm, security, or access controllers. They have their own resident software programs so that if the host computer malfunctions or the communications bus is severed, they will continue to operate, as programmed, in a stand-alone manner.

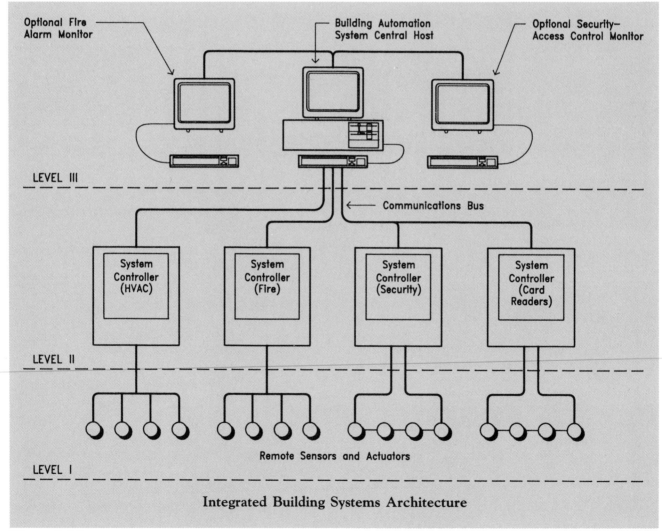

Integrated Building Systems Architecture

Figure 3.13

The information transmitted by the HVAC, fire, security, and access controllers to the central host can be segregated for monitoring and management purposes. The HVAC operator terminal and printer, therefore, can be located apart from the fire or security terminal and printer. In many buildings, operator terminals and printers are installed in both the security office and the maintenance office. Information transmitted by the controllers is prioritized as security or HVAC and displayed only at the appropriate location. Also, access by the operator's terminal to security and fire alarm controllers and point information can only be obtained from the security office, and access to HVAC information can only be obtained from the maintenance office. A third operator terminal could be installed at a selected location to receive all alarms and building management information. Management reports, such as energy audits, alarm logs, budget reports, and utility consumption reports, could also be generated at this location.

Although hardware and software information for building systems (HVAC, fire, security, access control, lighting) can be separated physically, they can also be integrated through interactive software programs. HVAC and lighting programs can interact for occupancy control; HVAC and fire alarm programs can work together for smoke evacuation, building evacuation, and stair tower pressurization: lighting and security programs can interact to automatically illuminate a zone when an intrusion alarm is initiated.

Part Two

Applications

The proper application of a Building Automation System requires careful planning as well as an understanding of what can be achieved by installing such a system. In an existing building with an old, obsolete HVAC control system, serious consideration should be given to the installation of a Direct Digital Control (DDC) system as a first step. The DDC system can act as a stand-alone control system, but can also be connected to a central computer in the future. The energy management programs available in the DDC software can provide some financial return to partially offset the initial investment. If central control and integration of other building systems (fire, security, access control, and lighting) are important, the central computer can be installed with the DDC system to accomplish these goals.

For new buildings, DDC is state-of-the-art for HVAC control systems. It provides flexibility, accuracy of control, and dependability. Connecting the DDC system to a central computer provides the option of integrating all building systems. It also offers the building owner and his operating staff numerous programs to assist them in monitoring and managing energy consumption and operating costs. The integration of building systems during the initial construction of a new building is the least expensive approach.

Chapter Four

Direct Digital Control (DDC)

For many years, computers have been used to monitor, supervise, and carry out energy management strategies on heating, ventilating, and air-conditioning systems. Pneumatic and electric control systems provided space temperature control and local loop control at the HVAC units. These control systems, although not energy efficient, did provide an acceptable level of comfort and, with a preventive maintenance program, a reasonable degree of reliability. The central computer, although expensive and complicated, gave building owners and operating personnel some management and control over energy-consuming systems in their buildings.

The advent of microelectronic technology revolutionized the automatic control industry. Microprocessor-based controllers could be installed in a mechanical equipment room and could perform the control loop functions previously assigned to conventional pneumatic, electric, and electronic control systems. They could also provide energy management and monitoring programs formerly performed by the central mainframe, or minicomputer. The introduction of the microprocessor reduced costs, increased reliability, improved accuracy, and provided a flexibility of control previously unachievable with conventional controls. Maintenance requirements were also dramatically reduced.

HVAC systems maintain comfort conditions in a building or building zone through the action of several *control loops*. A control loop consists of a sensor (for temperature, pressure, or humidity), a controller, and a final control element.

The sensor monitors the controlled variable and sends an input signal to the controller. The controller compares this signal to a preprogrammed set point. If the signal deviates from the set point, the controller sends an output signal to the final control element. The final control element then varies the controlled process to correct the deviation.

Input signals to the DDC controller take one of two forms: digital or analog. A digital input (DI) signal is a two-position signal, such as a switching action, where contacts are in either the closed or open position. This could be a signal from auxiliary contacts on a motor starter to indicate that the starter is energized. An analog input (AI) signal is a continuously changing input, such as a signal from a temperature or humidity sensor.

Output signals from a controller are also either digital or analog. A digital output (DO) signal is a two-position mode of control similar to a switching action. This mode of control would be used to start or stop a fan, or to control lighting in an on-off manner. An analog output (AO) signal is a variable output signal that can be used to modulate a valve or damper motor.

When all required remote sensors and controlled devices have been connected to the controller, the software programs for the appropriate control loops can be activated. Set points, reset schedules, sequencing, alarm limits, operating schedules, interlocks, and many other control options can now be programmed to create a fully operating control system. Modifications or revisions to control sequences, settings, or programs, can be made from an operator's terminal, without interfering with system operation.

Since energy management programs are also included in the controller memory, they too can be activated, if desired. In many cases, the sensors and the controlled devices connected to the controller for control loop purposes may be utilized for energy management programs as well.

The application of direct digital control to any system should be approached systematically. First, determine what control loops are required to create the desired environmental conditions. Include any energy management programs and monitoring requirements that you wish to accomplish. Next, make a schematic layout of the HVAC system and all its controlled devices, such as control valves, motorized dampers, magnetic starters, lighting contactors, relays, and so forth. Add sensors and other input devices in their appropriate locations. Extend all sensors, input devices, and controlled devices to a central location and label them digital inputs (DI), analog inputs (AI), digital outputs (DO), or analog outputs (AO).

Ventilation Control

DDC control loops for damper operation are available to provide ventilation requirements or to utilize outdoor air for cooling. The following are some of the more common programs.

Fixed Quantity of Outdoor Air

This control loop provides a fixed minimum amount of outside air for ventilation whenever the HVAC fan is energized. The minimum outdoor air damper moves to the full-open position when the fan is energized. When the fan is de-energized, the damper returns to the closed position. (See Figure 4.1)

Mixed Air Control

When the HVAC unit fan is energized, as sensed by a static pressure sensor, the damper control system becomes energized. A temperature sensor located in the mixed air modulates the outdoor and return air dampers to maintain mixed air temperature. A minimum outdoor air damper position can be set at the controller to ensure that minimum ventilation requirements will be met. If the system has an exhaust or relief damper, it will modulate in unison with the outdoor air damper. When the unit fan is de-engergized, the damper control system becomes de-energized, the outdoor air damper closes, and the return air damper opens. (See Figure 4.2)

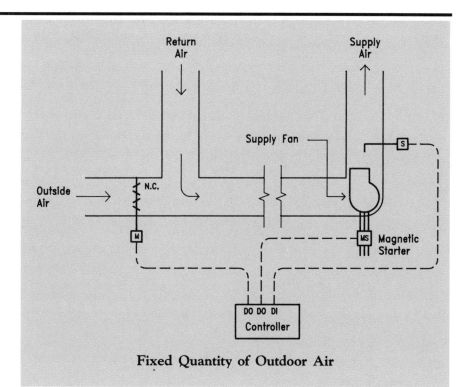

Fixed Quantity of Outdoor Air

Figure 4.1

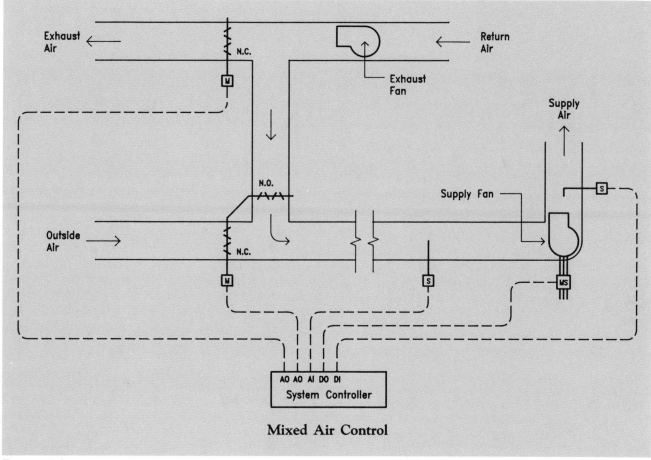

Mixed Air Control

Figure 4.2

Economizer Control of Mixed Air

The economizer control system provides a means of cooling a building using outdoor air, thereby saving energy that would otherwise be consumed if mechanical cooling were utilized. The basic concept for economizer control of mixed air is the same as for mixed air control (previously described), but also uses a system of dampers, temperature and humidity sensors, and actuators, to maximize the use of outdoor air for cooling.

A sensor mounted in the mixed air modulates the outdoor and return air dampers to maintain mixed air at a suitable temperature for cooling (usually 55°F). A minimum outdoor air damper position can be set at the controller to maintain minimum ventilation requirements. An outdoor air sensor returns the outdoor air damper to the minimum position when outdoor air temperature increases to a level at which it is no longer suitable for cooling (usually 65°F). If the system has an exhaust or relief damper, it will modulate in unison with the outdoor air damper to avoid overpressurization of the building. When the unit fan is de-energized, the outdoor air damper closes, and the return air damper opens.

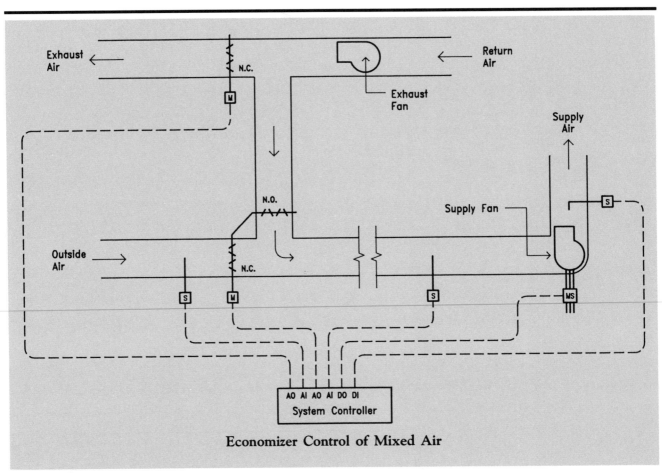

Economizer Control of Mixed Air

Figure 4.3

With sensors already installed in the mixed air and outdoor air, the controller can be programmed to reset the mixed air temperature at a higher level as outdoor air temperature decreases. This is an energy saving measure, since the mixed air must be heated before it is discharged into the building. Figure 4.3 shows the arrangement for either standard economizer or reset control.

Heating Control

There are numerous software strategies for heating applications. Programs are available to reset temperatures as a function of outdoor air temperature, for sequencing heating with cooling, or simply to control heat from discharge air or space temperature.

Constant Temperature Hot Water Control (Boiler)

A sensor installed in the system supply maintains water temperature by one of two methods: The controller can, through a digital output point, direct-fire the oil, gas, or electric source of heat. Or, the boiler can be maintained at a constant water temperature and a three-way control valve can be modulated, mixing boiler and return water to maintain supply water temperature. If the space heating units (radiation or heating coils) have two-way control valves, a differential bypass valve and sensor should be installed. As the two-way control valves close during low-load conditions, the pressure difference across the pump will increase. The differential pressure sensor will sense the increase and, through the controller, modulate the bypass valve toward the open position to maintain a constant pressure across the system. (See Figures 4.4 and 4.5)

Constant Temperature Hot Water Control (Heat Exchanger)

Constant temperature discharge control from a steam to hot water heat exchanger is accomplished by using a sensor in the discharge water that modulates a control valve in the steam supply. A normally-closed steam valve is recommended for this application. If the pump fails, the differential pressure sensor will sense the drop in pressure difference and close the steam valve. The differential pressure sensor will also maintain a constant pressure difference across the system by modulating the pump bypass valve as system two-way control valves close during load-low conditions. (See Figure 4.6)

Hot Water Reset Control (Boiler)

The addition of an outdoor air sensor to the control loop previously described for the constant temperature hot water control (boiler) will convey outdoor temperature information to the controller, which (through software programs) will then reset the discharge water temperature as a function of outdoor air temperature. This measure saves energy as well as wear on seats and discs of control valves located on radiation and heating coils. The outdoor air sensor can also shut down the system supply pump at a predetermined outdoor air temperature when heating is no longer required. (See Figures 4.7 and 4.8.)

Hot Water Reset Control (Heat Exchanger)

Again, the addition of an outdoor air sensor to the control loop previously described for constant temperature hot water control (boiler) will (through controller software programs) reset the discharge water temperature as a function of outdoor air temperature for the heat exchanger. This is an energy saving program and also saves wear on seats and discs of control valves located on radiation and heating coils. Again, the outdoor air sensor can shut down the system supply pump at a predetermined outdoor air temperature when heating is no longer required. (See Figure 4.9.)

Discharge Air Control (HVAC) – Heating

This control loop provides for a constant discharge air temperature. A duct-mounted sensor on the discharge side of the supply fan modulates a control valve in the steam or hot water supply to the heating coil to maintain the controller set point. If there is a chance that air could enter the heating coil at or below 32°F, a low-temperature safety controller should be installed on the discharge side of the heating coil. This controller, through software programs, can initiate corrective action if a coil freeze-up situation occurs. (See Figure 4.10.)

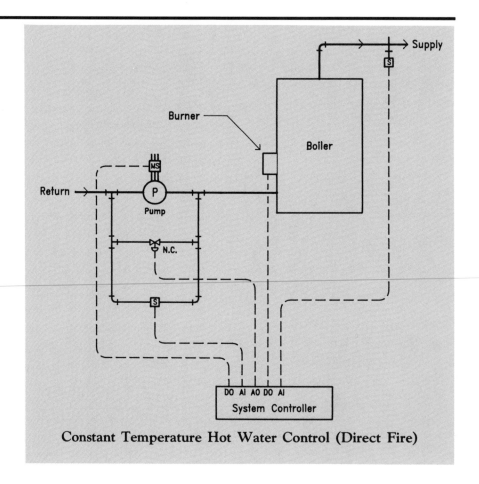

Constant Temperature Hot Water Control (Direct Fire)

Figure 4.4

If electric heat is utilized, the same control loop may be used except that the analog output to the control valve will be replaced with multiple digital outputs to control multiple stages of electric heat. If modulated electric heat is desired, an SCR (Silicon Control Rectifier) control may be installed, which can be controlled from an analog output similar to the control valve.

Discharge Air Reset Control (HVAC) – Heating

The addition of an outdoor air temperature sensor, operating in concert with a discharge air sensor, provides reset capability. The controller can receive signals from the two sensors and reset the discharge air sensor as a function of outdoor air temperature. (See Figure 4.11)

Space Temperature Control (HVAC) – Heating

A space temperature sensor acts as a primary sensor for an HVAC system. The sensor modulates a control valve in the steam or hot water supply to a heating coil, or cycles several stages of electric heat through

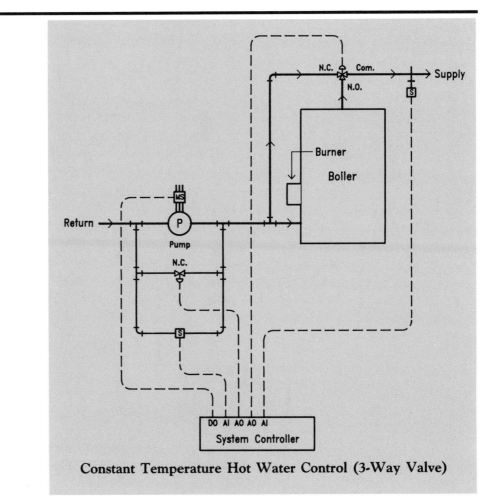

Constant Temperature Hot Water Control (3-Way Valve)

Figure 4.5

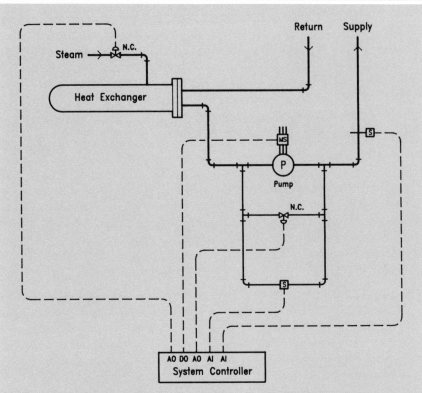

Constant Temperature Hot Water Control (Heat Exchanger)

Figure 4.6

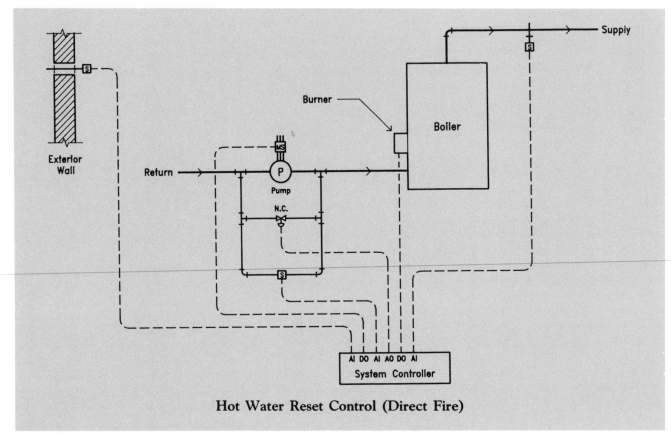

Hot Water Reset Control (Direct Fire)

Figure 4.7

multiple digital outputs. The discharge air sensor acts as a discharge low limit to maintain a minimum discharge air temperature. A reset software program can be applied so that the space sensor resets the discharge air temperature according to a predetermined schedule. Regardless of which strategy is applied, the layout will be as shown in Figure 4.12.

Cooling/Heating, Humidification/ Dehumidification Control

Included in Direct Digital Controllers are software programs for cooling control similar to those used for heating control, as discussed earlier in this chapter. The layout for discharge air control, in the case of a chilled water coil, is identical to Figure 4.10 in this chapter. If the source of cooling is a direct expansion coil, discharge control is not recommended. This could short cycle the compressor due to the large temperature drop when the compressor system is energized. The best application would be to control from a space sensor or a return air sensor. Discharge air reset control and space temperature control for chilled water cooling applications are the same as shown in Figures 4.11 and 4.12, respectively.

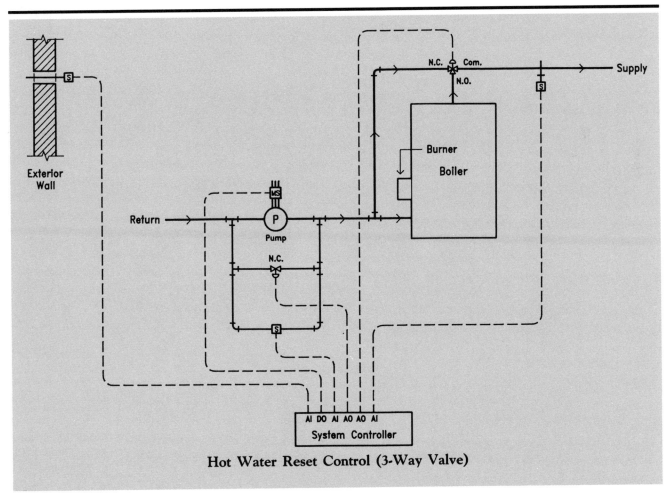

Hot Water Reset Control (3-Way Valve)

Figure 4.8

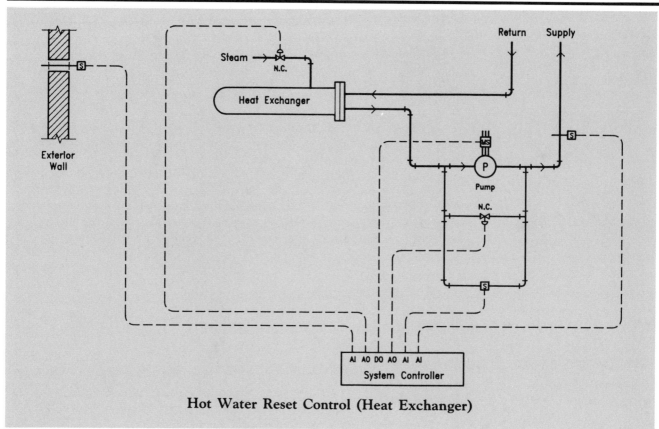

Hot Water Reset Control (Heat Exchanger)

Figure 4.9

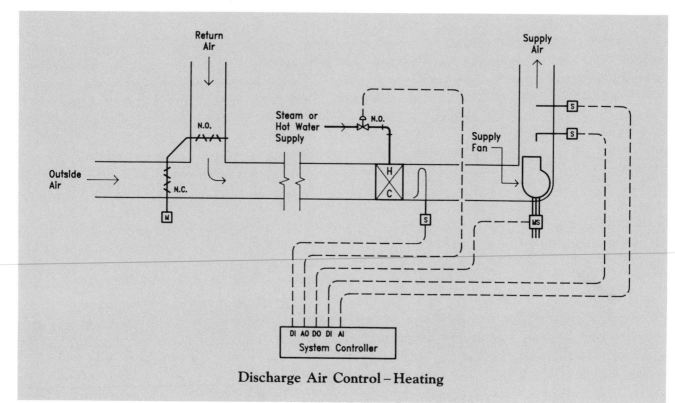

Discharge Air Control – Heating

Figure 4.10

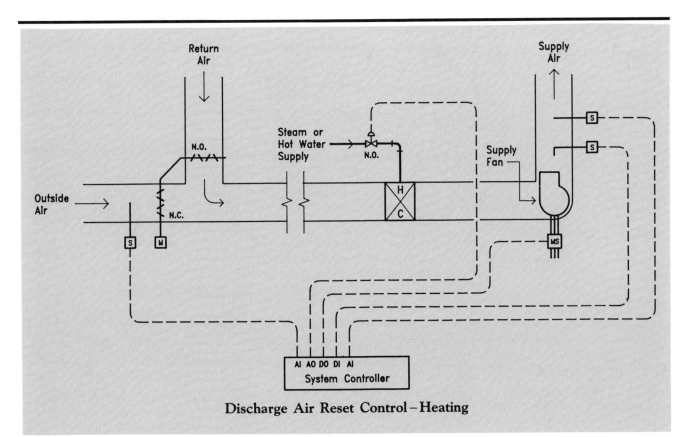

Discharge Air Reset Control – Heating

Figure 4.11

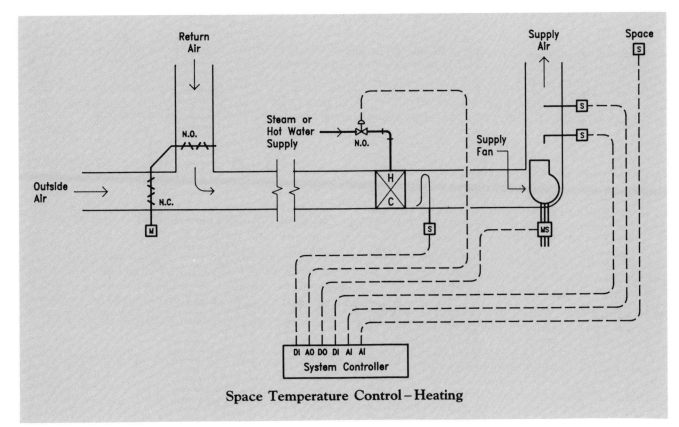

Space Temperature Control – Heating

Figure 4.12

Chilled Water Control (Centrifugal/Reciprocating)

Centrifugal refrigeration machines are controlled by a sensor installed in the system's chilled water supply. The sensor, through the controller, modulates the inlet vane actuator to maintain chiller discharge water temperature. The chilled water pump is started from a digital output point. A differential pressure sensor, installed across the pump, enables the chiller to operate through a digital output point when water flow has been established. An outdoor air sensor (dry bulb or dewpoint) or a return water sensor can be installed to reset the discharge water sensor setting. This reduces energy consumption by decreasing the pressure rise across the compressor.

Reciprocating chillers are also controlled by a sensor in the chilled water supply. However, the controller operates the compressors in a two-position manner through digital output points. Multi-stage chillers will have multiple digital output points. Like the centrifugal system, the chilled water pump for the reciprocating chiller is energized from a digital output point, and uses the same operating principles. Figure 4.13 shows the control loops for centrifugal and reciprocating chillers.

Dehumidification Control

Dehumidification control can be accomplished by means of a chilled water coil or a direct expansion (DX) coil. In the case of a chilled water coil, a space humidity sensor modulates, or two-positions, a control valve in the supply to the chilled water coil. A reheat coil should be installed downstream from the chilled water coil to reheat the dehumidified air. A space temperature sensor modulates the chilled water coil valve and reheat coil valve in sequence to maintain space temperature conditions. The space humidity sensor overrides the temperature sensor and opens the chilled water coil valve if space relative humidity exceeds its setting. The space temperature sensor maintains its set point by modulating the reheat coil valve.

If direct expansion dehumidification is used, the control sequence will be identical to that for chilled water, except that the compressor will be cycled through a digital output point. Figure 4.14 shows the layout for this application.

Humidification Control

A space humidity sensor maintains space relative humidity by modulating a control valve on a steam humidifier. A high-limit, duct-mounted humidity sensor located downstream from the humidifier overrides the space sensor and maintains 85% to 90% relative humidity in the supply duct. This prevents over-humidification and condensation in the supply duct. Some humidifiers require a two-position mode of control which would necessitate a digital output point. When the unit supply fan is de-energized, as sensed by a static pressure sensor, the humidifier becomes inoperable. The humidification control loop is shown in Figure 4.15.

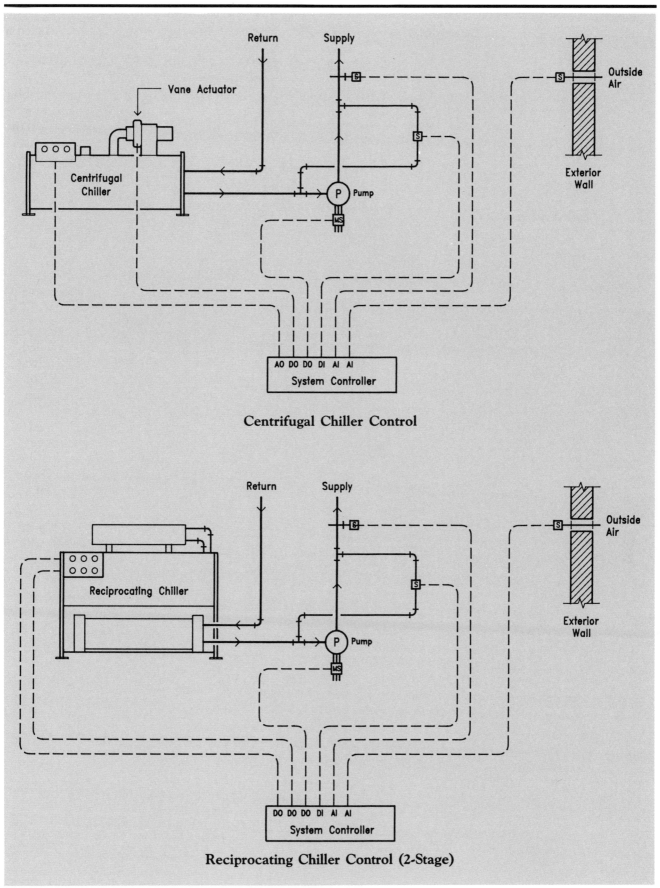

Centrifugal Chiller Control

Reciprocating Chiller Control (2-Stage)

Figure 4.13

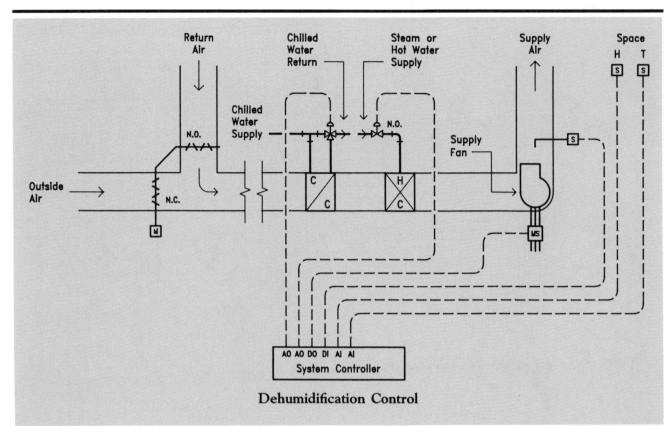

Dehumidification Control

Figure 4.14

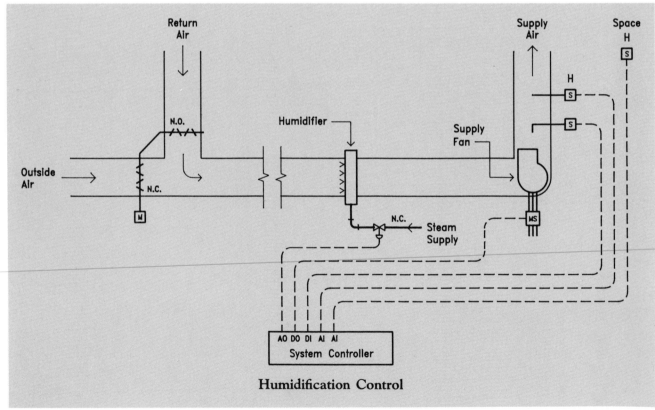

Humidification Control

Figure 4.15

Heating – Cooling Sequencing

A space temperature sensor, or duct-mounted discharge air sensor may be used to control the heating coil valve in sequence with the chilled water coil valve. As an energy saving feature, most DDC controllers have a *deadband program*. This means that an adjustable number of degrees can be programmed between the heating and cooling modes. During this deadband, both the heating and cooling coil valves will be closed, and no energy will be consumed. Although this will result in wider-than-normal temperature swings in the space, these swings can be limited to acceptable comfort levels. Figure 4.16 shows the heating-cooling sequencing control loop, with hot water or steam for heating and chilled water for cooling.

When direct expansion cooling and electric heat are controlled, the same sequencing control software is used. However, the heating and cooling control outputs will change to digital outputs, as required by the number of stages of electric heat and direct expansion cooling. When the supply fan is de-energized, as sensed by the static pressure sensor, the direct expansion cooling and the electric heat will be inoperable.

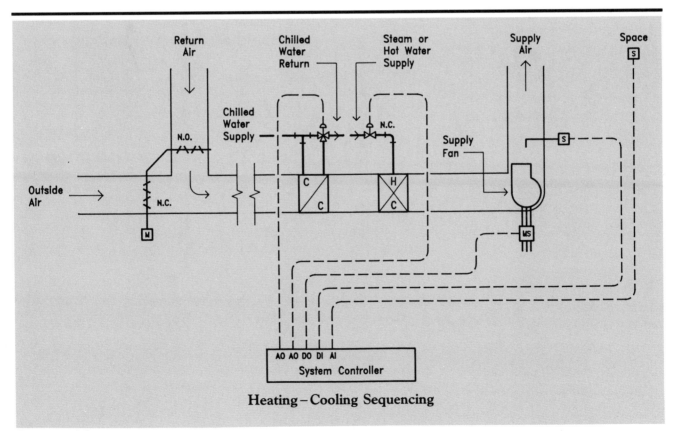

Heating – Cooling Sequencing

Figure 4.16

Humidification – Dehumidification Sequencing

This control loop employs the same basic software programming as the heating-cooling sequencing. A space humidity sensor modulates the humidifier steam valve in sequence with the chilled water valve to maintain space relative humidity. As mentioned under *Dehumidification Control*, a reheat coil is necessary to prevent subcooling during the dehumidification cycle. The space temperature sensor controls the heating coil valve and the cooling coil valve, in sequence, to maintain space temperature conditions. Figure 4.17 displays the layout for this control system.

Static Pressure Control

Static pressure can be controlled by two methods: 1) by modulating fan inlet vane dampers, or 2) through variable speed drives on the fan motor. In variable air volume (VAV) systems, the static pressure sensor is located in the main supply duct to the terminal units. The sensor, through the controller, maintains supply duct static pressure by modulating the system fan inlet vanes, or by modulating the fan speed through a variable speed drive. (Figure 4.18 shows the layout for controlling inlet vanes to maintain duct static pressure in a variable air volume system.)

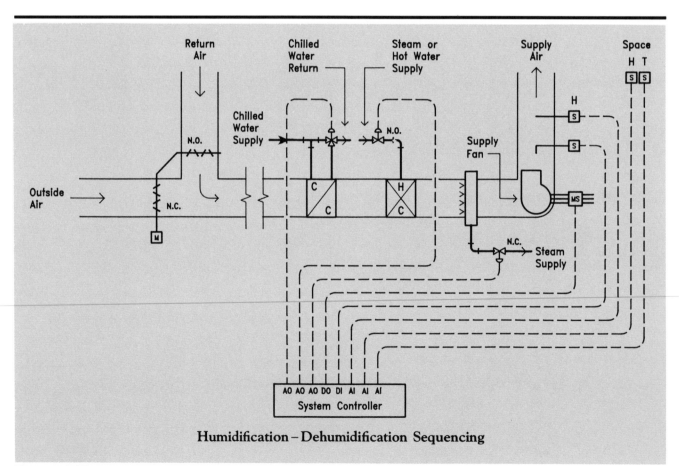

Humidification – Dehumidification Sequencing

Figure 4.17

Variable speed drives provide excellent fan speed control and are usually designed for 3 to 15 pounds, 4 to 20 milliamps, or 2 to 12 VDC inputs. These inputs fall within the scope of most DDC controller output capabilities. For direct building or floor static pressure control, the sensor is located in the space and is referenced to an appropriate location outside the building. With the VAV system, the controller either modulates the return air fan inlet vanes or controls a variable speed drive to maintain static pressure in the space.

Variable Air Volume System Terminal Box Control

Most variable air volume systems supply air at approximately 55°F year-round. With a decrease in space temperature, the space sensor, through the zone controller, throttles the VAV box, thereby reducing the volume of cool air entering the space and allowing the space temperature to increase. Areas located around the perimeter of the building may require reheat coils or radiation to maintain space temperature during severe winter periods.

VAV systems require some form of static pressure control at the fan. This is necessary to compensate for periods in which a large number of VAV box dampers in the system move to the closed position, resulting in an increase in static pressure in the supply duct. See the section titled *Static Pressure Control* and Figure 4.18 in this chapter for an explanation of this control loop.

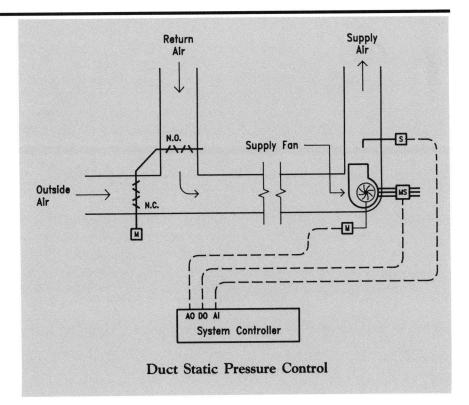

Duct Static Pressure Control

Figure 4.18

Pressure Independent – Cooling Only

In cooling-only applications, the space sensor (through the controller) regulates the airflow regardless of variations in supply duct pressure. The minimum and maximum airflow that are needed (as sensed by an airflow sensor) to ensure ventilation requirements are met during low-load conditions and prevent excessive delivery of cool air during periods of high-demand. Figure 4.19 shows this application.

Cooling and Electric Reheat

When electric reheat is included in the terminal box, an adjustable deadband can be programmed between the cooling and heating cycles. On a drop in space temperature, the sensor modulates the damper toward the preset minimum position. The damper will remain at the minimum position throughout the desired deadband. On a continued drop in space temperature, the electric heat will be energized in stages. (See Figure 4.20.)

Cooling and Hot Water Reheat

This application is identical to that outlined in the section *Cooling and Electric Reheat*, except that the stages of electric heat are replaced with a modulating valve controlling a hot water coil. The adjustable deadband also applies to this application. (See Figure 4.21.)

Fan-Powered Return Air Heat

In this application, the cooling volume damper is sequenced with the return air fan. Upon a decrease in space temperature, the space sensor, through the zone controller, modulates the cooling damper to the minimum position. The damper will remain at this minimum position throughout the adjustable deadband. Upon a continued drop in space temperature, the return air fan will be energized. (See Figure 4.22.)

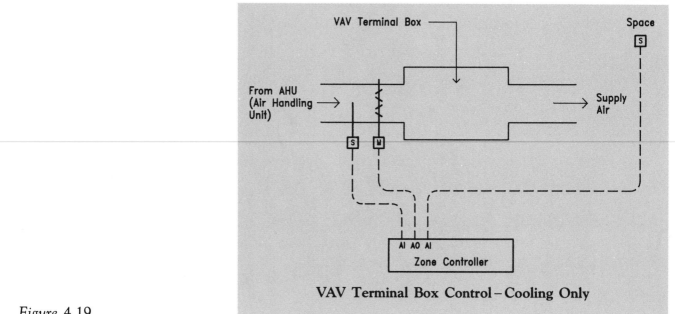

VAV Terminal Box Control – Cooling Only

Figure 4.19

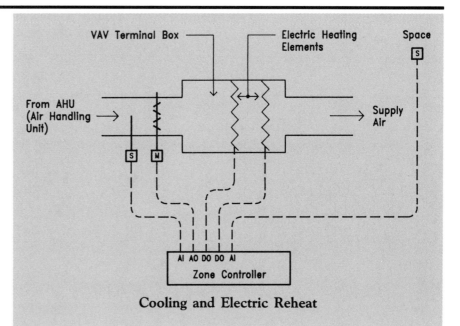

Cooling and Electric Reheat

Figure 4.20

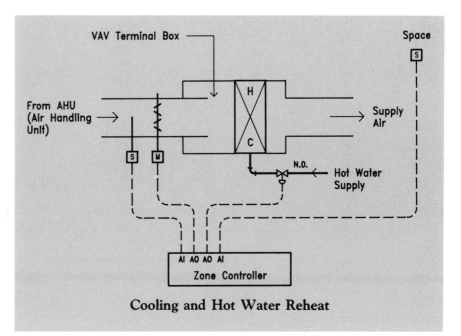

Cooling and Hot Water Reheat

Figure 4.21

Fan-Powered Return Air and Electric Heat

In this application, the electric heat stages and the return air fan are sequenced with the cooling damper. Upon a drop in space temperature, the space sensor causes the cooling damper to modulate to the minimum position. The damper will continue at the minimum position as long as conditions remain within the adjustable deadband range. With a continued drop in space temperature, the damper will adjust to a new heating minimum position, and the return air fan will be energized. On a call for additional heat, the stages of electric heat will be cycled on in sequence. This is shown in Figure 4.23.

Fan-Powered Return Air and Hot Water Heat

This application is identical to that outlined in the section *Fan-Powered Return Air and Electric Heat*, except that the stages of electric heat are replaced with a modulating valve-controlled hot water coil. The adjustable deadband applies to this application as well. (See Figure 4.24.)

Night Temperature Setback/Setup and Morning Warm-Up/Cool-Down for Variable Air Volume Systems

When a single HVAC unit supplies multiple zones, night temperature setback or setup is traditionally accomplished by cycling the unit fan from a centrally located night thermostat. Morning warm-up or cool-down is traditionally controlled from a space thermostat or return air controller. If the zones have different occupancy schedules, energy is wasted under this method of control. This is because the system operates as one large zone, resulting in the heating or cooling of unoccupied zones.

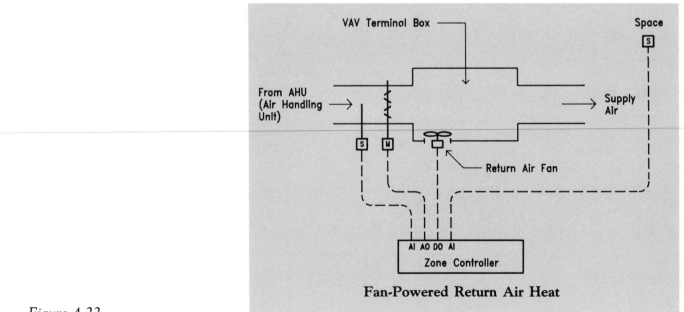

Fan-Powered Return Air Heat

Figure 4.22

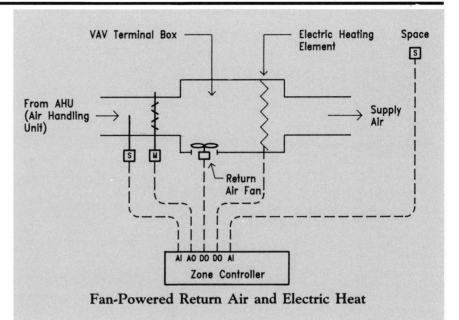

Fan-Powered Return Air and Electric Heat

Figure 4.23

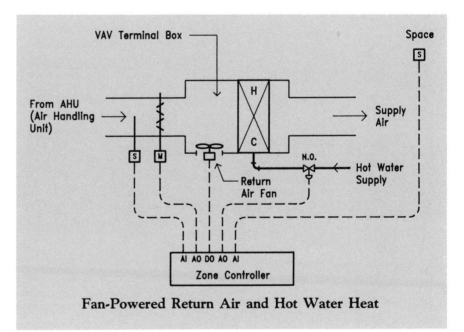

Fan-Powered Return Air and Hot Water Heat

Figure 4.24

With direct digital control, the VAV controller for each zone can be programmed for the appropriate occupied/unoccupied schedule and unoccupied setback/setup temperatures. Another DDC benefit is the ability to provide warm-up or cool-down on a zone-by-zone basis. When a zone controller switches from the unoccupied to the occupied mode, the VAV box will switch to the full heat or full cooling position until the space temperature reaches the occupied setting of the space sensor. Control will then revert to the space sensor.

A Typical HVAC Unit DDC System

Figure 4.25 depicts a DDC System applied to a typical heating, ventilating, and air-conditioning unit. The control system is made up of several control loops discussed in this chapter: economizer control of mixed air, heating-cooling sequencing, and humidification-dehumidification sequencing. The next step is to activate the appropriate software programs to accomplish the required sequence of operation as follows.

When the unit fan is energized, as sensed by a static pressure sensor in the supply duct, the damper control system becomes activated. A mixed air sensor maintains mixed air temperature (55°F) by modulating the outdoor air, return air, and exhaust dampers. When outdoor air temperature exceeds the setting of the outdoor air sensor, the outdoor and exhaust air dampers return to the minimum open position as programmed at the controller. The return air damper takes the corresponding open position.

A space temperature sensor, through the controller, maintains the space temperature by modulating the heating coil valve in sequence with the chilled water coil valve. A space humidity sensor, through the controller, maintains space humidity. Upon a drop in space relative humidity, the humidifier steam valve modulates toward the open position, subject to a duct-mounted high-limit humidity sensor. With a rise in space relative humidity, the humidifier steam valve modulates to the closed position, followed by the opening of the chilled water coil valve to provide dehumidification. During the dehumidification cycle, the space temperature sensor modulates the heating coil valve to maintain space temperature conditions.

A low-temperature controller, with its capillary located on the discharge side of the heating coil, will, through the controller, de-energize the unit fan, close the outdoor and exhaust dampers, and open the return damper if discharge air temperature drops below its setting (40°F). Whenever the unit fan is de-energized, as sensed by the supply duct static pressure sensor, the damper control system will be de-activated, closing the outdoor and exhaust dampers and the humidifier steam control valve.

Host Computer

All DDC controllers are capable of operating in either a stand-alone mode or as part of a distributed process system with other controllers. Controllers can share data with each other, or with a central host computer, over a common communications bus. To assist with stand-alone operation, or to simplify maintenance, most system manufacturers offer a portable, hand-held operator's terminal which allows direct access to system point data and operating information through any controller. This terminal can also be used to issue commands or alter programs.

The central host computer is a PC that acts as an operator-machine interface to the DDC controllers. The controllers provide local loop control and energy management functions over HVAC systems. The host computer hardware includes the personal computer, with hard disk and diskette drives, keyboards, CRTs, and printers. Figure 4.26 shows a typical BAS with DDC.

Most manufacturers' software provides menu-driven selections of system functions. They provide conversational guidance in data file programming; English language descriptors and messages; current status of controller inputs and outputs; historical log printouts; and alarm monitoring, override control and programming of controllers linked to the PC over the communications bus. Figure 4.27 shows a typical CRT screen format. The printer maintains hard copy documentation of alarms and logs. Some more common application programs are described in the following sections.

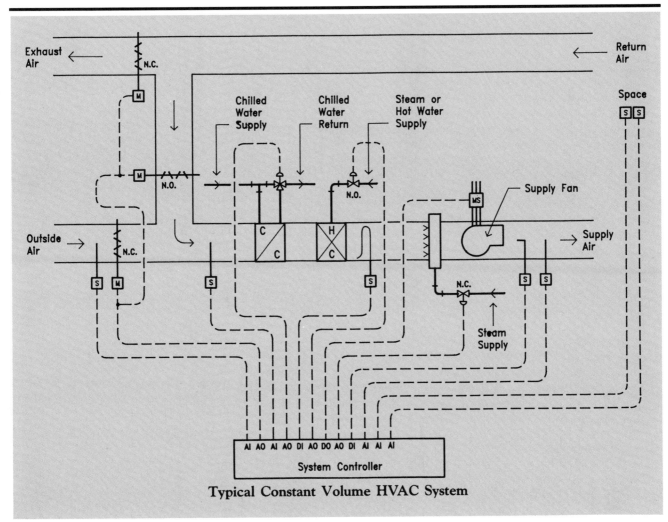

Typical Constant Volume HVAC System

Figure 4.25

Storage and Retrieval of Historical Data

This capability can assist operating personnel in forecasting breakdowns by observing trends. Operators can initiate the storage of information on any point. They can also initiate reports manually, or program a weekly or calendar schedule for automatic initiation of reports containing the desired point data, including alarms, return-to-normal, or status of selected points.

Action Messages

Action messages can be programmed to display on the CRT and print whenever a specified point generates an alarm or trouble condition. This is extremely helpful to an operator monitoring critical points or systems. It is also helpful to a new operator who is unfamiliar with emergency procedures.

Energy Auditor Program

The energy auditor program allows the user to keep records of building energy consumption. This data assists operations personnel when reviewing and evaluating current and historical energy consumption trends. A spreadsheet format is utilized to convert measured data into

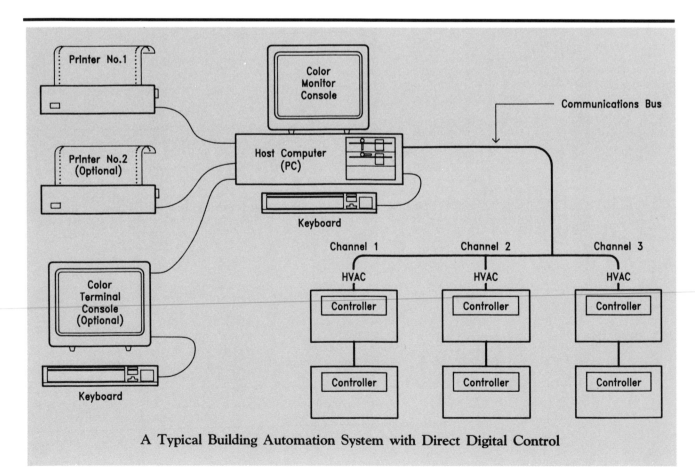

A Typical Building Automation System with Direct Digital Control

Figure 4.26

meaningful statistics and historical comparisons. Output can be a listing of energy consumed per day (kilowatt hours, cubic feet of gas, gallons of oil), or the measured information can be converted to management reports (kilowatt hours or BTUs per square foot of building). The user can design the report format: everything from labels and descriptive messages, to content and display format. This program makes it possible for building management personnel to predict building energy trends, evaluate energy conservation strategies, anticipate problems, and budget energy costs more effectively.

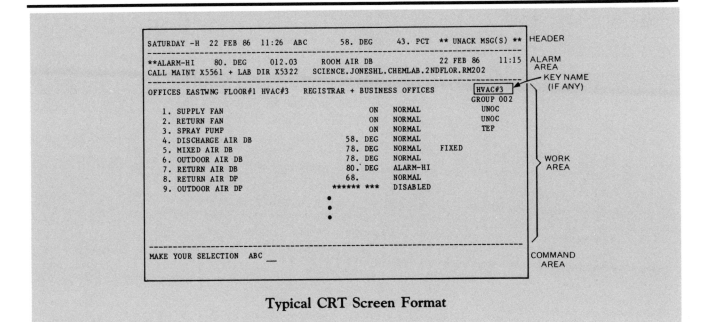

Typical CRT Screen Format

Figure 4.27

Maintenance Management Program

This program allows the user to schedule preventive maintenance based on the run time of the equipment. Printouts document the preventive maintenance tasks, the materials and skills required to accomplish the task, and any special tools or instruments needed. The preventive maintenance activities may be scheduled based on calendar time (every 3 months, for example) or on the run time of the equipment (for example, every 400 hours). Filter changes can be scheduled upon the generation of a dirty filter alarm. This is an event occurrence method of scheduling filter changes. Work orders for preventive maintenance can be printed with the following information:

- Maintenance task instructions
- Equipment to be serviced within the building and location
- Type of tradesman required (electrician, pipefitter, etc.)
- Time required to accomplish the task
- Materials and/or special tools required.

Upon completion of the task, the actual time and materials used are entered into the data files. The program updates costs and tracks maintenance costs on an ongoing basis. This information can be retrieved and printed out for historical information on maintenance costs.

To assist in the supervision of maintenance personnel, a printout of incomplete work orders can be generated. Maintenance activity reports and work orders can be printed manually by the operator to cover unusual situations. If a historical cost summary is required on a piece of equipment, the program can print a complete maintenance activity report on that piece of equipment over any period of time.

Since the data stored in this program is sensitive, most systems have more than one level of operator access. A lower level operator may be cleared to access completed work orders and request management reports, while a higher level operator can perform the lower level functions as well as change database information.

Color Graphics

Most systems on the market today have a color graphics software option. A library of standard symbols and icon-based graphics makes on-line creation of graphics possible. Every input-output point in the DDC controllers can be displayed graphically at the host CRT. Floor layouts, zones, building elevations, HVAC systems, and piping layouts can be shown. Point data on the graphic display can include point status or value, color-coded to indicate normal or alarm condition. Representation of point status or value can be in animated form. For instance, fan status can be shown with the blades rotating when the fan is energized, and stationary when the fan is off. Damper blades can actually be shown changing from the open to the closed position. When a point goes into alarm, the associated graphics will be displayed, giving the operator a visual location of the point. The color graphics option provides a high-resolution display of building information, reports the status of all system points, and displays critical alarms for operator acknowledgment. All displays appear in multiple colors for clarification and distinction of information.

Group and Historical Logs

From the operator's terminal, an operator can request an *all-points summary*, an *alarm summary*, or a *group point summary*. An all-points summary generates a printout of the status or value of every point in the system. The alarm summary generates a printout of all points currently in alarm. The group point summary produces a printout of the status or value of all points in a logical group.

Historical trend logs store the status and value of selected points on a hard disk. This can be done over designated periods of time. It is a valuable tool in the hands of maintenance personnel seeking to identify potential HVAC system problems.

Line Graphs, Bar Charts and Pie Charts

This program takes historical database information and transfers it to a spreadsheet program. The operator can construct bar or pie charts or line graphs using the spreadsheet information. These charts and curves display data and trends in more understandable form than a simple column of figures. Figure 4.28 shows some sample charts and graphs.

Voice Communications Option

Most BAS's on the market today offer a voice communications option. This program allows the user to receive notification of alarms or other specified information at a preprogrammed telephone number during predefined periods of time. When an alarm or critical condition occurs at one of the predefined points, the software automatically dials an assigned telephone number or its backup. A synthesized voice message is issued to the person who answers. This feature ensures notification of critical alarms to authorized personnel during periods when a building is unoccupied. Authorized personnel can call the host computer from an off-site touchtone telephone and request the status of predefined points. Predefined points can also be issued commands if deemed necessary. Figure 4.29 illustrates the voice communications option.

With direct digital control, control systems can be designed with state-of-the-art accuracy, dependability, and flexibility. Building owners and their operating staff can take advantage of a multitude of computerized programs to assist them in running a more efficient, comfortable and safe building.

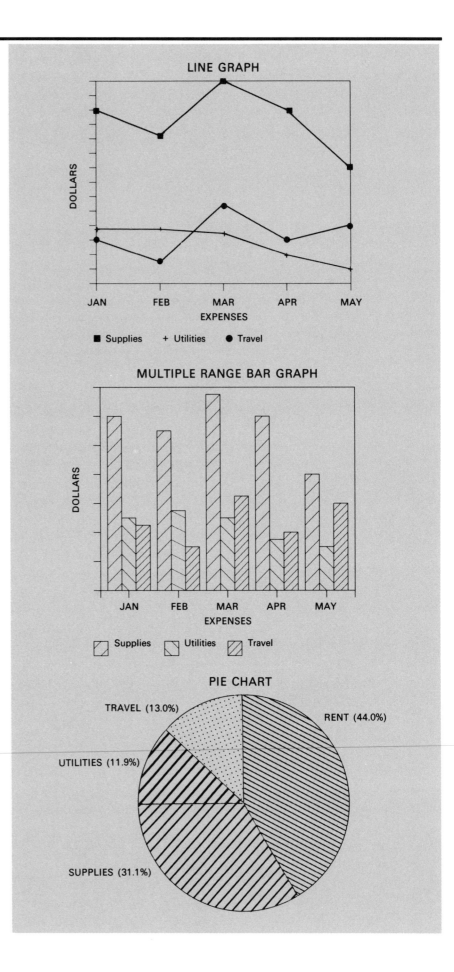

Figure 4.28

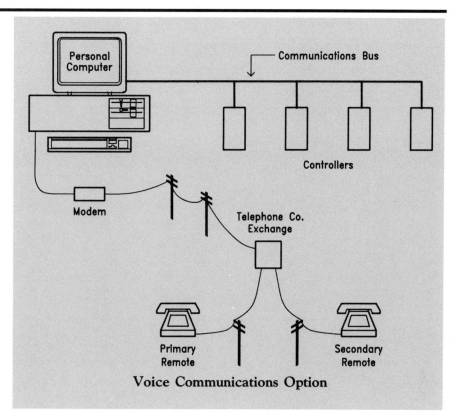

Figure 4.29

Chapter Five

Energy Management Programs

Before installing a Direct Digital Control system, whether the system is a new building installation or replacing an existing conventional control system, consideration should be given to the activation of energy management programs. These programs are incorporated in the controller, and can generate impressive financial paybacks that may partially offset the installation cost. Most major manufacturers of these systems will provide an estimate of energy savings and submit, with their proposal, a payback summary which may be achieved by instituting various energy management programs.

Approximately 67% of the energy consumed in an office building is electrical. Thirty-three percent is oil, gas, and purchased steam or hot water. A breakdown of the electrical energy consumed reveals that cooling systems use 40%, lighting systems use 33%, heating, 12%, and the remaining 15% is consumed by other means. The energy management strategies, resident in the DDC controllers, address all of these areas of energy consumption.

A major source of energy savings, in most buildings, is through reduced run time of HVAC systems. Electricity is saved by reducing fan and/or pump operations. Additional savings are achieved through reduced outdoor air intake, thereby relieving a burden on the heating and cooling equipment. Some energy management programs use outdoor air for cooling. This reduces the load on the mechanical cooling equipment, resulting in reduced energy consumption. Reset programs can maintain the temperature of air delivered to the spaces at optimum levels to satisfy comfort conditions. This eliminates the energy waste that results from overheating and subcooling.

Electrical power companies are promoting programs to reduce the demand for electricity and relieve pressure on their limited generating capacity. Many companies offer rebate incentives to encourage customers to install energy management systems that will reduce building electrical demand. For example, the replacement of inefficient light fixtures and bulbs can result in savings on electrical costs. Additional savings can be realized by reducing lighting levels, and turning off lights during unoccupied periods.

All energy management programs should be approached with a degree of caution. If not properly applied, these programs could damage expensive equipment, compromise comfort conditions, and have a negative impact on air quality. What follows is a discussion of some of the energy management programs used today.

Duty Cycle Program

The *duty cycle program* reduces electrical energy consumed by the fan in an HVAC system by cycling the fan on and off. The off periods are a function of space temperature, as sensed by a space temperature sensor. When space temperature is at the midpoint of a preprogrammed comfort range, the fan can be turned off for an extended period of time without affecting comfort. As space temperature approaches the end of the comfort range, the off periods are decreased (Figure 5.1).

The operator programs the duty cycle period, the comfort range in degrees Fahrenheit (for example, 68°F-78°F), minimum on and off times, and the maximum off time. These times are set to avoid the equipment damage that can result from too rapid cycling. When the space temperature is at the midpoint of the comfort range, as sensed by a space temperature sensor, loads such as fans are turned off and outdoor air dampers are closed for the maximum off time in the period. When space temperature deviates from the midpoint, the off periods shorten proportionally. At the extreme ends of the comfort range, which usually occur at outdoor design conditions, the duty cycle program cancels and the load is energized continuously.

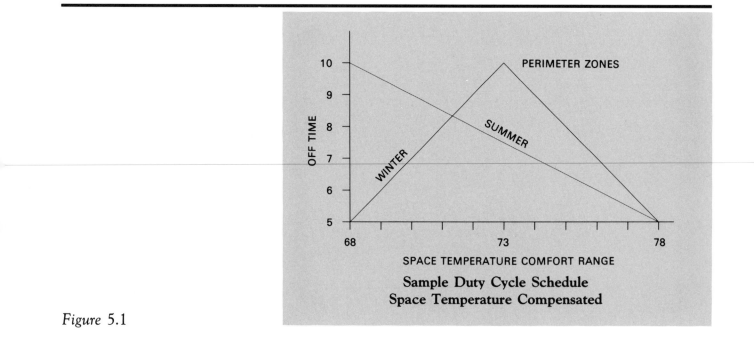

Sample Duty Cycle Schedule
Space Temperature Compensated

Figure 5.1

Multiple space sensors may be used for space temperature compensated duty cycling programs. In this case, the program selects the highest space temperature sensed for cooling and the lowest temperature sensed for heating. When multiple electrical loads are duty cycled, the program automatically adjusts the cycle periods so that they do not occur simultaneously, thereby causing a power surge. This is shown in Figure 5.2.

There are certain cautions that should be exercised in the application of the duty cycle program. First, the best candidates for this program are heating, ventilating, and air conditioning unit fans (under 100 horsepower only), exhaust fans, and hot water pumps. HVAC units and exhaust fans serving critical areas such as computer rooms, clean rooms, laboratories, operating rooms, and areas where toxic materials or plastics of any kind are stored, should not be duty cycled.

Second, many questions have been raised about the effect of duty cycling on electric motors. Research has shown that the only serious problem that may arise is overheating of the motor during startup. The two important factors to consider when duty cycling electric motors are the minimum time between starts (minimum interval) and the minimum off time. Figure 5.3 depicts guidelines for cycling induction motors.

The minimum off time is determined by the cooling requirements of the motor after running at its normal operating temperature. For small motors, three minutes is adequate cooling time before restarting. Notice in Figure 5.3 that the interval and the off time are extended as the motor size increases. As a general rule, motors larger than 100 horsepower should not be duty cycled. If HVAC fan motors of this size are installed, alternatives should be considered, such as cycling inlet vanes or mixing dampers, to achieve energy savings. However, belts and magnetic starters, if properly sized and installed, will be minimally affected by a properly applied duty cycle program.

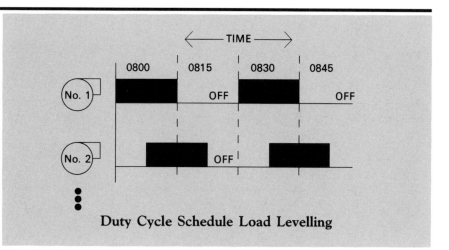

Duty Cycle Schedule Load Levelling

Figure 5.2

Power Demand Limiting Program

Electrical utility companies charge for energy consumption over an established billing period (usually one month). These charges cover fuel costs, operating costs, and the utility's investment in generating equipment.

The charges are calculated based on readings taken from the building's electric meter. Every commercial building has a kilowatt-hour meter. This meter records, on a continuous basis, the kilowatt-hours of electricity consumed by the building. At the end of the billing period, the utility company reads the meter and bills the customer for the energy consumed. The charges are frequently based on a sliding scale, so that the more energy consumed, the lower the unit cost. Added to the cost per kilowatt-hour is a fuel adjustment charge to cover the fluctuating cost of fuel used to produce electricity during that billing period. This is a fixed charge per kilowatt-hour regardless of the number of kilowatt-hours consumed.

Since the demand for electricity by commercial customers is heavier at certain times of the day than at others, the utilities must have equipment and transmission lines capable of handling these peaks. Although these peak demand periods occur only occasionally and for relatively short periods of time, the generating capability must nonetheless be available. *Demand charges* were established to pay for the installation and maintenance of this excess generating capability. Figure 5.4 shows a typical building load profile.

Demand charges are based on the average kilowatts consumed during a demand interval. A *demand interval* is a specific period of time established by the utility company. This period of time varies depending on the utility company, but in most cases is 15 or 30 minutes. Demand charges are based on the maximum demand (demand peak) during any demand interval in the billing period. Many utility

Motor Cycling Guidelines			Nema Design Types A–F	
HP	1/4–10	10–20	20–50	50–100*
Min Interval (Minutes)	10	20	30	40
Min Off Time (Minutes)	3	5	7	7
Suggested Max Off Time	5	7	10	10
Min % KW Savings	30%	25%	23%	17%

*With reduced voltage start only

Figure 5.3

companies have a *ratchet clause* which states that the customer must pay for this maximum demand for eleven months. If, during the eleven-month period, a new, higher demand peak occurs, this will establish the basis for the demand charge over the next eleven months.

The *power demand limiting program* monitors electrical consumption during each and every demand interval, and sheds (turns off) assigned loads as required to reduce demand. Before such a program can be implemented, however, the building owner or engineer must determine which loads in the building are "sheddable" for brief periods of time without creating serious environmental problems. For instance, shedding air-conditioning for a computer room or an operating room would not be acceptable, nor would shutting down a ventilating unit for hospital patient areas or intensive care units. Air-conditioning units and exhaust fans supplying general office areas, on the other hand, can be shed for brief periods, as can units supplying conference rooms, lobbies, coffee shops, and cafeterias. Also, the set point of large, electrically driven chillers can be set up a few degrees as a temporary demand limiting strategy.

Loads are prioritized according to their importance to environment and comfort conditions in the building. Low-priority loads are shed first and restored last. High-priority are the last to be shed and the first to be restored. If multi-speed motors are assigned to the load shed program, they can be shed in multiple steps.

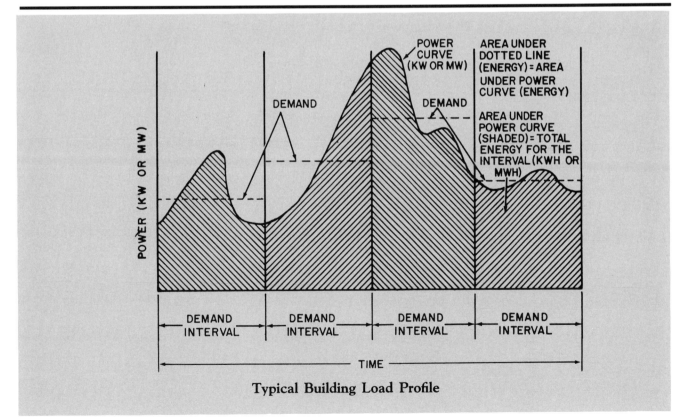

Typical Building Load Profile

Figure 5.4

The power demand limiting program monitors power consumption from the data input of a pulse transmitter located on the building's electric meter. However, not all building meters are equipped with a pulse transmitter. The building owner should check with the power company before applying this program to determine whether or not his building has one. If the transmitter does not exist, there will be a charge by the power company for the installation.

The power demand limiting program during the 15- or 30-minute demand interval, computes the energy consumed and forecasts the power demand at the end of the interval. After completing the forecast, the program adds or sheds loads to maintain the demand level within programmed limits.

If all assigned loads have been shed and electrical consumption continues to increase, the program automatically shifts the demand limit upward. When seasonal variations in electrical demand occur, the program compensates rather than maintain the original limits. Figure 5.5 is a sketch showing how the basic demand limiting program works.

Unoccupied Period Program

With the exception of residential-type buildings, most buildings have occupied and unoccupied periods. Office buildings are typically occupied 40 to 48 hours per week, which means that they are unoccupied approximately 72% of the time. Hospitals have certain areas, such as administrative offices, gift shops, cafeterias, x-ray departments, and operating rooms which are not occupied full-time. If unoccupied buildings or building zones are heated, cooled, or ventilated to the same levels as when occupied, energy is being wasted.

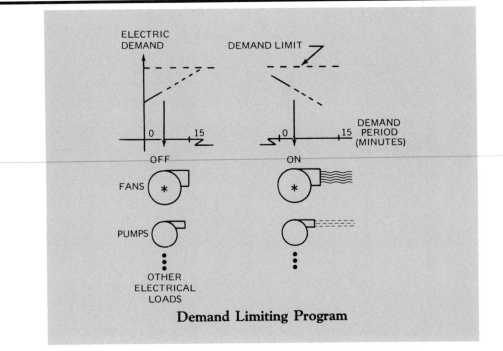

Demand Limiting Program

Figure 5.5

Considerable energy savings can be achieved by setting temperatures back during unoccupied periods in the heating season, and up during unoccupied periods in the cooling season. Savings on hot water radiation systems can be achieved by decreasing water temperatures during the unoccupied periods. Air-handling systems, which are cycled to maintain reduced temperatures during unoccupied periods, should have their fresh air dampers closed.

The unoccupied period program, or night cycle program, is primarily a heating season function. It can, however, maintain a high space temperature limit during the cooling season, if desired. A space temperature sensor maintains space temperature at preset levels during unoccupied periods by cycling the heating or cooling source. If multiple sensors are used, the program will control from the lowest temperature in the heating season and the highest temperature in the cooling season.

If humidity is of critical importance in a building or building zone, a relative humidity sensor can be installed to override the program and cycle the air-handling system to maintain a high or low space relative humidity level.

Optimum Start-Stop Program

Program clocks have been used for many years to automatically determine the occupied/unoccupied periods for buildings or building zones. These devices have saved building owners millions of dollars in energy costs at a minimal installation cost. In order to achieve comfortable building temperature at occupancy time, however, the clock must be set for outdoor design heating or cooling conditions. For instance, if it takes two hours to bring a building up to comfort levels at occupancy, under design heating conditions, the program clock must be set to turn the heating equipment on two hours prior to occupancy. At the end of the occupied period, the program clock is usually set to place the building in the unoccupied mode. However, such design conditions occur only a few days each winter.

During non-design days, heating equipment that starts two hours early results in wasted energy. The same principle applies to the cooling season.

The optimum start-stop program is an adaptive energy-saving program that uses intelligence and the flywheel effect (energy retention capacity) of a building to save a considerable amount of energy beyond that which can be saved with the program clock. The *optimum start program* monitors space and outdoor air temperatures several hours prior to the programmed occupancy time of the building or zone. If the space temperature is within comfort limits, the heating or cooling equipment will be started exactly at building or zone occupancy time. If space temperature is not within comfort limits, the program will calculate the correct start-up time necessary to achieve comfortable temperatures at occupancy time. After weekends or holiday periods, a building may require longer warm-up or cool-down periods. The optimum start program automatically compensates by extending the run time of the heating or cooling equipment prior to occupancy.

Figure 5.6 is a graphic representation of the optimum start program for both the summer and winter modes.

Figure 5.7 shows a typical optimum stop graphic representation for the heating mode.

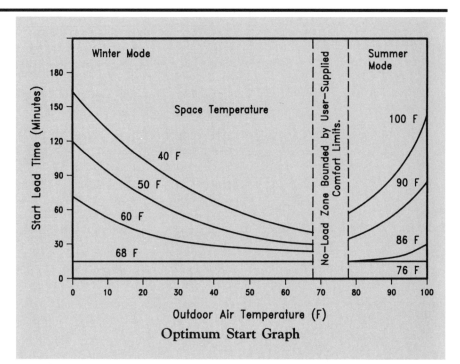

Figure 5.6

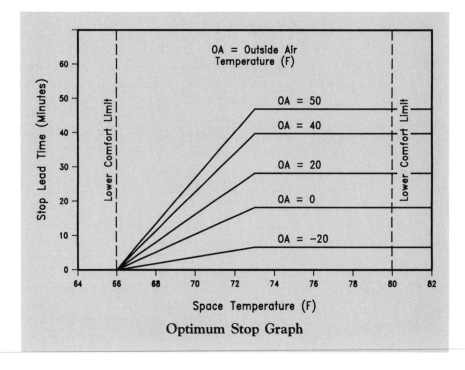

Figure 5.7

The *optimum stop program* occurs at the end of the building or zone occupancy time. It calculates the earliest time during the occupancy period, that the heating or cooling equipment can be shut down, allowing the flywheel effect of the building to stabilize temperature levels until the end of the period.

Both optimum start and optimum stop programs can function with multiple space sensors. The program will be sensitive to the area or zone sensor which is exposed to the most extreme temperatures.

Unoccupied Night Purge Program

During the summer cooling season it is not unusual for the outdoor air temperature to drop considerably at night. Frequently, during the early morning hours prior to building occupancy time, the outdoor air temperature is below building space temperatures. This cool outdoor air can be utilized to cool the building, thereby eliminating the need for mechanical cooling during early morning occupancy hours. This free cooling will generate energy savings and also save wear on the mechanical cooling equipment.

At a preprogrammed time in the early morning hours, the program begins to monitor space and outdoor temperature and humidity. If space conditions indicate a need for cooling, and if outdoor air conditions are suitable, the *Night Purge Program* is initiated. The program starts the HVAC supply fan and associated exhaust fan, and opens the outdoor air damper 100%. Warm air from the building continues to be purged until the space temperature and relative humidity indoors reach the same levels as the outdoor air conditions, or until the morning start-up program begins. The outdoor air temperature must be above a preselected minimum to ensure that the program is operable only during the cooling season. Figure 5.8 shows a flow chart of the night purge program operation.

The *night purge program* can be applied to most HVAC systems that are capable of using 100% outdoor air. Some package-type HVAC units and rooftop units are limited mechanically to admit 10 or 20% outdoor air, and therefore do not qualify.

Enthalpy Program

The traditional mixed air economizer control system saves energy by using outdoor air for cooling rather than mechanical cooling equipment. This control loop functions by sensing dry bulb temperatures. When the outdoor air dry bulb temperature reaches a level at which it is no longer suitable for cooling, the fresh air damper closes to the minimum position and the mechanical refrigeration equipment becomes operable. It is at this point that the *enthalpy control system* becomes important.

The economizer system, since it measures dry bulb temperature only, keeps the system on return air throughout the cooling season. The enthalpy control system, however, selects the air source (return air or outdoor air) which has the lowest total heat *(enthalpy)*, and therefore requires the least amount of heat removed by the cooling coil.

The enthalpy program monitors the temperature and relative humidity or dewpoint of the outdoor and return air. During occupied periods when cooling equipment is in operation, the program calculates the total heat, or enthalpy, of the outdoor air and return air. It then positions the outdoor air and return air dampers to use the air source with the lowest total heat or least enthalpy.

This program will work with HVAC systems that have sprayed or unsprayed cooling coils. The system must have the capability of operating on 100% outdoor air, or 100% return air.

Load Reset Program

HVAC systems are sized for design conditions, or the peak load during the heating and cooling seasons. These peak load periods are of short duration and occur infrequently, which means that most systems operate under partial load conditions most of the year. This results in inefficient performance. If the systems are not properly controlled, energy is wasted.

The *load reset program* controls heating and/or cooling to maintain comfort conditions in the building while consuming a minimum amount of energy. This is accomplished by resetting the heating and cooling discharge air temperatures to satisfy the zone with the greatest heating or cooling load. When no heating or cooling is required, the program de-energizes both sources.

If an HVAC system supplies multiple zones with different exposures such as North, South, or locations (interior, exterior), the heating and cooling requirements could vary widely. The load reset program will accept and evaluate inputs from multiple zone sensors and issue signals, if necessary, to actuate the heating, cooling, or heating *and* cooling systems. If the lowest zone temperature is below the preprogrammed comfort level, the heating source will be enabled and the discharge air temperature will be reset in proportion to the amount of deviation from the comfort level. If the lowest zone temperature is above the preprogrammed comfort level, the heating source will be shut off.

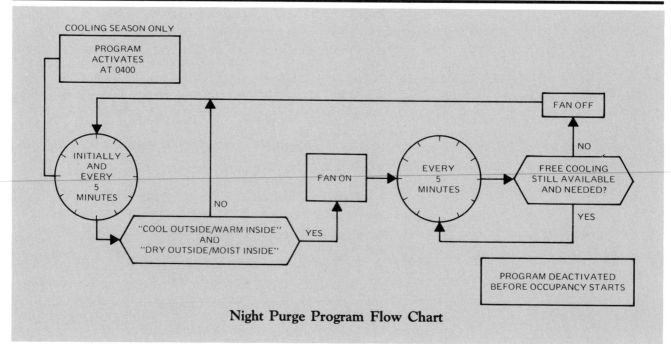

Night Purge Program Flow Chart

Figure 5.8

If the highest zone temperature is above the preprogrammed comfort level, the cooling source will be enabled and the discharge air temperature will be reset in proportion to the amount of deviation from the comfort level. If the highest zone temperature is below the preprogrammed comfort level, the cooling source will be shut off.

The load reset program also has a high-limit humidity or dewpoint option. If space relative humidity is critical, space relative humidity or dewpoint sensors can be installed. Any one of these sensors can lower the cooling reset schedule to compensate for increasing space relative humidity levels.

In a terminal reheat system, the HVAC unit discharge air temperature will be reset to the point at which the zone sensor with the greatest demand for cooling is satisfied. In dual duct and multi-zone units, the cold deck is reset according to the zone sensor with the greatest demand for cooling, and the hot deck is reset according to the sensor with the greatest demand for heating. When all space sensors are satisfied at the preprogrammed comfort levels, both the heating and cooling sources will be shut off.

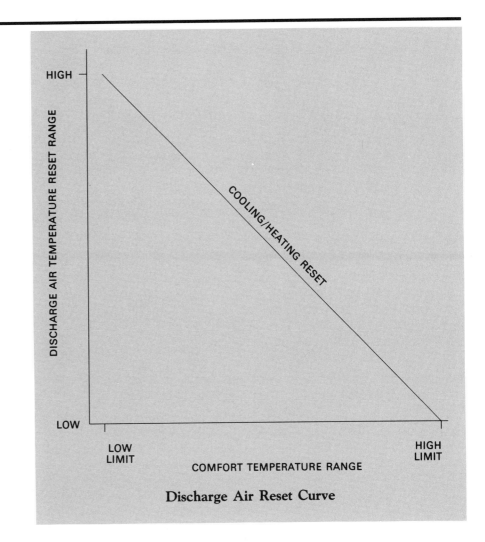

Discharge Air Reset Curve

Figure 5.9

Figure 5.9 shows how discharge air temperature in a hot and cold deck system is reset as a function of the sensor with the greatest demand for heating or cooling.

Zero-Energy Band Program

This program finds its application in HVAC systems that have heating and cooling capability. The program saves energy by avoiding simultaneous heating and cooling of air delivered to spaces. The space comfort range is divided into three sections: heating, cooling, and zero-energy band. In the zero-energy band portion of the comfort range, both heating and cooling sources are disabled. When space temperatures drop to the lower end of the comfort range, the heating source is energized and the discharge air temperature is reset upward. When space temperature enters the upper end of the comfort range, the cooling source is energized and the discharge air temperature is reset downward. The reset action that takes place in the heating and cooling sections of the comfort range is a preprogrammed function of the temperature range of the heating or cooling section. In other words, a load reset program operates in both the heating and cooling sections of the comfort range (See Figure 5.10).

When an HVAC system serves different zones, or rooms, the heating and cooling requirements will vary widely. Therefore, when the temperature in the coldest room or zone falls into the low end of the comfort range (the heating section), heat will be delivered. The temperature of the air, however, will be reset according to the degree of penetration into the heating section. Cooling follows a similar process. When the temperature in the warmest room or zone rises into the high end of the comfort range (the cooling section), cool air will be delivered. The temperature of the air will be reset according to the cooling load in the room or zone. When all room or zone temperatures are within the zero-energy band, both heating and cooling sources will be disabled. Within the zero-energy band, the program can reset discharge air temperature by modulating the mixing dampers (outdoor air and return air) to maintain space comfort conditions.

Host Computer

In a distributed process system, the remote DDC controllers perform their energy management and control loop tasks on a stand-alone basis. The central host computer, however, acts as the operator-machine interface to the controllers and remote sensors. This is accomplished over a communications bus, which links the controllers to the host.

The host computer is an intelligent operator-machine interface which consists of hard diskette drives, keyboard, CRT, and printer. The hard disk holds all operating programs as well as data files for the PC host and controllers. The diskette drive provides the capability for storage of historical data and logs.

The keyboard and CRT provide the primary operator interface with the system. The PC keyboard is similar to a typewriter keyboard and includes a numeric keypad. The CRT displays conversational menus which guide the operator to available functions. When a programming task is to be performed, the operator receives guidance on what data to enter, through conversational prompting and fill-in-the-blanks requests.

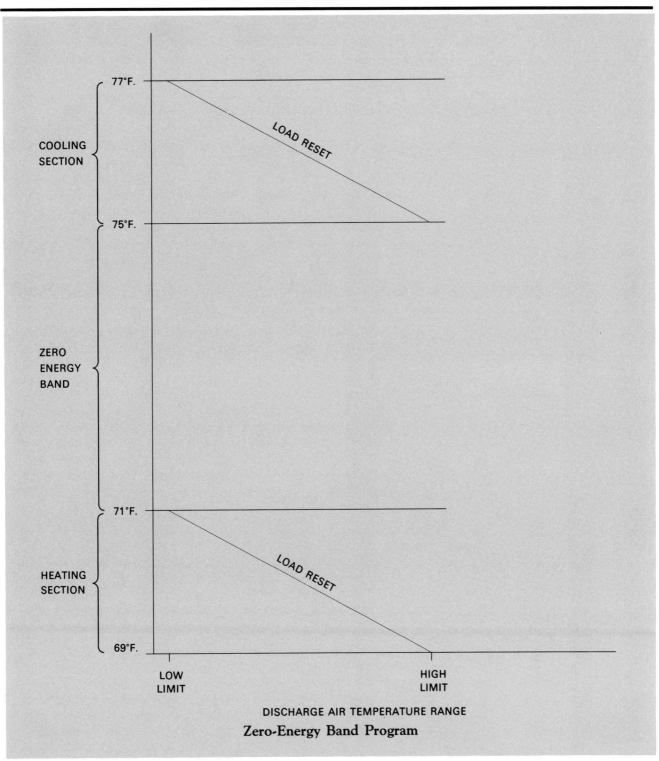

Figure 5.10

The printer operates in a receive-only mode and provides hard copy records of alarms and logs. All logs, such as all points, alarms, trends, and groups can be requested from a log menu and printed for record purposes. Multiple printers may be installed so that alarms and logs may be segregated.

The host software provides menu-guided or system selection functions. It also provides conversational directions in data file programming, English language descriptors and action messages, current status and historical logs, alarm monitoring, programming, and program override of all controllers on the communications bus.

The central host computer develops and prints management reports based on energy consumption information supplied to it by remote sensors. The programs discussed in the following sections assist facilities management personnel in evaluating the effectiveness of the energy management programs used in their buildings. If the programs are not accomplishing the anticipated results, changes in parameters and strategies can be made through the central host.

Energy Auditor Program

The energy auditor program can convert measured data, such as electric demand meter input information, into valuable statistics and trend data. A printout can be generated listing peak demand for each day of the month. Inputs from electric and gas meters can supply data to develop energy consumption trends in spreadsheet format. The output can be a simple listing of data, or a calculation can convert the data to management-level reports such as KWH per square foot, or cubic feet of gas per square foot. This program can be a valuable tool to building owners and managers for tracking costs, predicting trends, budgeting, and managing energy consumption.

Graphs and Charts

Line graphs and bar or pie charts can be displayed on the CRT to depict trends in any of the variables supplied for the spreadsheet program. Through the trend log spreadsheet programs, bars or graphs can be constructed to display data in a more easily understood form. Trends in daily peak demand, or kilowatt-hour consumption, are easier to read and comprehend when represented as bars of various heights, as opposed to multiple columns of figures. Figure 5.11 shows sample charts and graphs.

Action Messages

Action messages can be programmed to appear on the CRT and to be printed whenever a specified point generates an alar or trouble condition. This feature is extremely helpful to an operator monitoring critical points or systems. Action messages can be programmed to interact with the power demand program. The operator can be notified if all available loads have been shed and a new demand peak is imminent. The CRT and printer can display corrective action instructions.

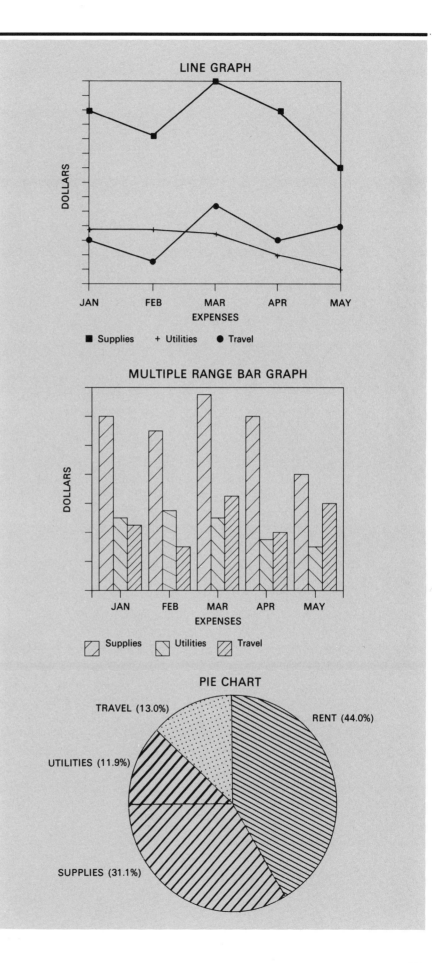

Figure 5.11

All of the energy management programs outlined in this chapter are included in the DDC controller software. As energy costs escalate and energy shortages occur, these programs offer an answer to the control and management of building operating costs.

Assigned personnel can receive alarms at authorized telephone numbers during predefined periods of the day or night. When an off-normal condition occurs on one or more specified points, the software automatically dials the preprogrammed telephone number or its alternate and issues the appropriate voice message. The program then issues a request for commands.

An authorized operator can dial into the host computer from an off-site location and request the status of a point or issue commands. Specific status points and command points can be assigned and programmed for access through the dial-in program. This program enables building operations personnel to remain in contact with their building after regular business hours.

Chapter Six

Lighting Control

Lighting accounts for 30-40% of the electrical operating costs in commercial buildings. Yet the average commercial building is unoccupied approximately 75% of the time. In many of these buildings the lights remain on during the unoccupied periods. During occupied periods, the lighting levels in these buildings are often considerably higher than required by the occupants for the work they perform. Furthermore, the heat produced by lights in a building, although beneficial during the heating season, places a burden on the air-conditioning systems during the cooling season.

There are steps that can be taken to reduce lighting costs while maintaining optimum lighting levels during occupancy periods in large buildings. Building owners can institute lighting control programs which effectively operate the lighting system in their building by means of manual and automatic control. Because lighting requirements vary according to the tasks being performed and the occupancy schedule of a particular zone, lighting control programs provide the flexibility of controlling illumination levels according to occupant needs and occupancy schedules. For example, lights can be turned on in the cafeteria zone in the morning and maintained at required levels for the breakfast period. At the end of this period, the levels can be reduced slightly until lunch, when they will be restored. Following lunch the light levels will decrease. At the end of the working day the lights can be turned off or maintained at reduced levels for security purposes. Lighting can be preprogrammed for the entire work week, weekends, and holidays.

In order to apply lighting control successfully, however, the building's lighting system must meet certain requirements. Distribution panels must contain lighting circuits only – no outlets, or motors. If a mixture of lighting, outlets, motors, and other electrical equipment exists on a panel, the lighting must be transferred to a separate circuit breaker panel. There is no need for local switching of lights under a centralized lighting control system. Wall-mounted light switches in individual offices or classrooms are not necessary.

Lighting control generally works best on open area lighting design such as in cafeterias and open landscape office areas. The availability of natural light is also important, to supplement artificial lighting, thereby reducing electrical energy costs.

Building Automation Systems can provide two major strategies for lighting control. One strategy is a simple on-off program for occupied-unoccupied control. The second strategy monitors total lighting levels in the perimeter zones of the building and supplements the natural light with artificial light to maintain the required lighting levels.

Occupied-Unoccupied Lighting Control

The first strategy to reduce light operating costs is called *occupied-unoccupied lighting control*. This lighting control system is a time-based program that schedules the on/off time of lights, for a building or zone to coincide with occupancy schedules. Multiple programs are available, which makes it possible to arrange for lighting to match occupancy schedules of different zones in a building. Although this is a very basic method of lighting control, it can generate impressive savings in electrical costs if no lighting control program is presently in operation.

Lighting Level Control

Another way to reduce the costs associated with lighting is to control the *level* of lighting in a building or building zone. Lighting level control is accomplished by two different methods: *multi-level lighting* and *modulated lighting*. Before applying either of these programs, however, a study should be conducted of building lighting requirements. Zones with similar lighting requirements should be listed. All activities in a zone should have approximately the same occupancy schedules. Window and wall configurations should be similar throughout a zone. Exposure to daylight should be uniform throughout the zone. Zones should be small enough so that one or two people working overtime will not require excessive lighting.

Multi-Level Lighting Control

Multiple levels of illumination are made possible by dividing three- and four-lamp fixtures into three and four circuits respectively. By switching the ballasts, three-lamp fluorescent fixtures can provide four levels of control: off, one-third, two-thirds, or full-on lighting. Figure 6.1 shows four levels of lighting control. Two- or four-lamp fixtures can provide three levels of control: off, one-half, or full-on lighting.

Multi-level lighting is applied primarily to perimeter zones where daylight is available to supplement the artificial light. Light sensors are installed in these zones to energize fixture lamps in an on-off manner. The sensors measure levels of natural light in the zone and, through the lighting program, energize the appropriate number of fixture lamps to maintain preset lighting levels.

Cleaning crews and security and maintenance personnel frequently work after normal building occupancy hours. A program can be developed to provide the required light levels to accommodate cleaning schedules for each zone. Another option is to install occupancy sensors to energize lights when employees enter a zone after normal occupancy hours. The program will turn off the lights after a preprogrammed amount of time has elapsed. The central host computer can also be used to manually override lighting programs. If there is a 24-hour security guard station in the building, specific zones can be illuminated from the terminal at the guard station for unscheduled activities.

Modulated Lighting Control

Modulated lighting control differs from multi-level control in that it is not limited to two or three levels of lighting, but can modulate to any level over the complete range from off to full-on.

Modulated lighting is specifically designed to control the standard four foot, rapid-start fluorescent lamps found in most large buildings today. Standard fluorescent lamps, when first installed, provide more light output than required, so that rated output is provided over the life of the lamp. This type of lamp consumes 100% of its power requirement over its entire life. The modulated lighting program provides even light levels throughout the lamp life by adjusting output to the needs of the zone occupants. Figure 6.2 shows the comparison of standard lighting levels and power consumption to that of modulated lighting control.

Modulated lighting control systems employ the same basic closed-loop process used to control temperature and humidity through DDC systems. Light level sensors containing photocells are mounted on acoustic ceiling panels or on standard electrical boxes. The sensors send input signals to a zone control unit in direct proportion to the ambient

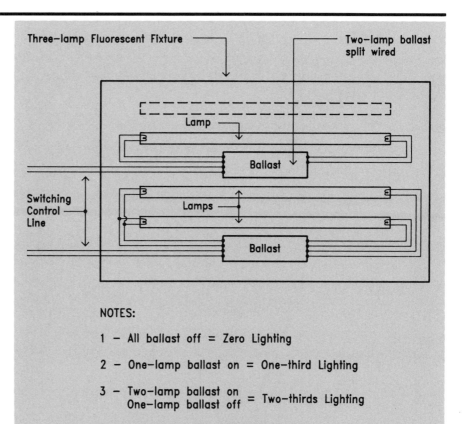

NOTES:

1 – All ballast off = Zero Lighting

2 – One-lamp ballast on = One-third Lighting

3 – Two-lamp ballast on
     One-lamp ballast off = Two-thirds Lighting

4 – All ballast on = Full-on Lighting

Four Levels of Lighting Control

Figure 6.1

light levels. The zone control unit has an illumination level set point. The unit compares this set point to the input signals from the illumination sensors. If ambient light levels deviate from this set point, the zone controller sends an output signal to the power controller. The power controller, in turn, adjusts the power output level delivered to the lighting control ballasts. (Keep in mind that because the power controller is installed ahead of the circuit breaker panel, it is imperative that the breaker panel contain only lighting circuits and no wall outlets or appliances.) The control ballast then adjusts the output of the rapid-start fluorescent lamps to compensate for the change in illumination.

Control ballasts are specifically designed for lighting control systems. A full-power control ballast is used where 100% lighting capacity is required. Reduced power control ballasts may be used for applications in which lighting is overdesigned, or for lighting circuits near windows, where 100% light output is not required.

Figure 6.3 shows a closed loop, modulated lighting block diagram.

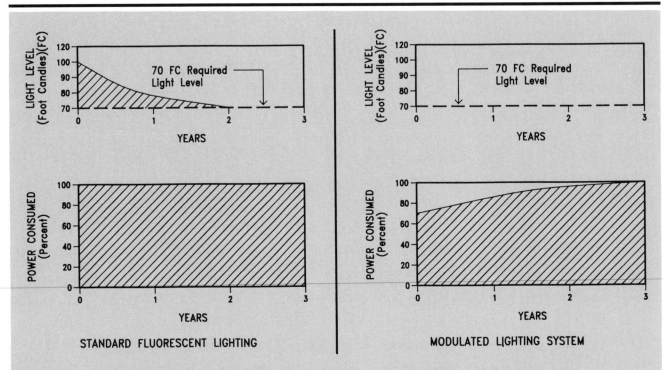

Modulated Lighting vs. Standard Lighting Levels and Power Consumption

Figure 6.2

Host Computer

In lighting control systems, the host computer acts as the operator-machine interface to the control and energy management programs. Through the color graphic software package, color-coded floor layouts can be constructed. The light status (on-off) of a room, a zone, or an entire floor can be displayed graphically on the CRT. If lighting levels in a zone drop below or exceed preset levels, an alarm will be activated at the host computer and the appropriate floor will appear on the screen to show the operator which zone is in alarm. From the operator's terminal, the operator can override lighting programs to accommodate unusual occupancy schedules. This can be done weeks in advance and can be programmed to occur on specific nights and for specific time periods. The printer will provide hard copy documentation of the program override date, zone, and length of time. The computer can be programmed to calculate the total kilowatt-hours consumed during the override period for tenant billing purposes. It is also possible to combine lighting with HVAC systems for occupied-unoccupied scheduling.

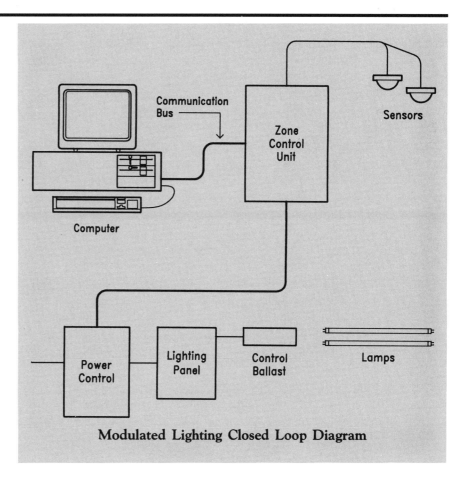

Modulated Lighting Closed Loop Diagram

Figure 6.3

Voice Communications Option

Most Building Automation Systems on the market today offer a voice communications option. Authorized personnel and/or tenants are issued system access password codes. During off hours, they can call the host computer from an off-site touchtone telephone and command the turning on of lights and, in some cases, the appropriate HVAC systems for a specific area over a specific time period. When occupants arrive, the appropriate area lighting will be on and/or the associated HVAC systems will have pre-conditioned the area to comfort levels. Figure 6.4 illustrates the voice communications option.

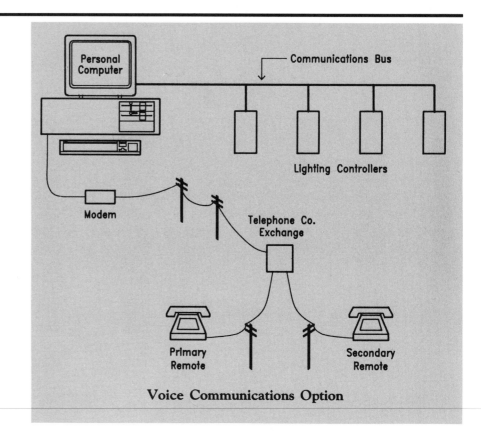

Voice Communications Option

Figure 6.4

Chapter Seven

Fire Safety Integration

Fires occur in buildings of all types and sizes. As a result, life and property protection have become a top priority for insurance companies, government agencies, and numerous private fire protection organizations. Most states, cities, and towns in the United States have strict building codes which specify minimum standards for design and construction. These codes were written specifically to minimize the risk of fire in public buildings as well as protect the lives of the building occupants. One system that has become a standard fixture in most buildings today is a fire alarm system.

Fire alarm systems fall into two general categories: *life safety* and *property protection*. Life safety systems are installed in buildings that are "occupant-intensive," such as schools, health care facilities (hospitals, nursing homes, mental institutions), prisons, and multiple-story office buildings. The primary goal of a life safety system is early detection and prompt notification, to allow for an orderly and safe evacuation of the building occupants.

Property protection systems are installed in buildings that contain valuable merchandise or confidential or irreplaceable records and documents. Computer rooms, warehouses, museums, art galleries, and libraries fall into this category. The primary goal of a property protection system is detection of fire in the early stages of combustion. Prompt notification of fire fighting personnel and strict guidelines regarding fire extinguishing methods are of major importance in such buildings. See Figure 7.1.

Basic Fire Alarm Systems

First and foremost, the fire alarm system must have the ability to detect a fire and announce its existence and location to the proper authorities at a continuously occupied, central site. There are a number of functions, either directly or indirectly related to fire management, that a fire alarm system can perform. For example, fire alarm systems may also contain and/or evacuate smoke, and release secured doors to allow people to vacate the building.

All fire detection and alarm systems contain basic components. Every fire alarm system must have:

- an initiating circuit
- an indicating circuit
- a control unit

The *initiating circuit*, sometimes called a *sensing circuit* or *sensing loop*, is a circuit designed to detect a fire condition. The *indicating circuit*, sometimes called the *bell loop*, provides notification of the fire condition. The fire alarm control is described in detail later in this chapter. Each of these components is shown in Figure 7.2 and is discussed in the sections that follow.

Initiating Circuit

The detection of a fire condition originates with a sensor or sensors in the fire alarm system initiating circuit. Each type of sensor has unique characteristics that make it better suited for certain applications. (The National Fire Protection Association pamphlet titled "NFPA 72E Sensors" contains standards for the installation of fire alarm system sensors.) Some types of sensors commonly found in initiating circuits are:

- thermal detectors
- smoke detectors
- sprinkler heads (waterflow)
- manual pull stations

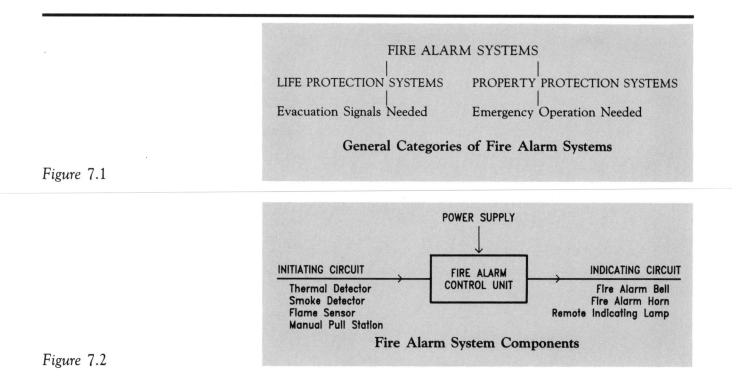

Figure 7.1

FIRE ALARM SYSTEMS

LIFE PROTECTION SYSTEMS   PROPERTY PROTECTION SYSTEMS

Evacuation Signals Needed   Emergency Operation Needed

General Categories of Fire Alarm Systems

Figure 7.2

POWER SUPPLY

INITIATING CIRCUIT
Thermal Detector
Smoke Detector
Flame Sensor
Manual Pull Station

FIRE ALARM CONTROL UNIT

INDICATING CIRCUIT
Fire Alarm Bell
Fire Alarm Horn
Remote Indicating Lamp

Fire Alarm System Components

Thermal Detectors

Thermal detectors, as the name implies, are actuated by temperature. These detectors are usually mounted on the ceiling in the area to be protected. The time required for a thermal detector to respond to a fire is directly related to its proximity to the fire.

There are basically two types of thermal detectors: *fixed temperature* and combination *rate-of-rise/fixed temperature*. Fixed temperature detectors are available with different, fixed actuation temperatures. The most common is 135°F. They are also available for high-temperature actuation (190°, 200°F). See Figure 7.3.

Combination rate-of-rise/fixed temperature detectors have two different methods of actuation. They will operate at a specific temperature, in the same manner and with the same actuation temperatures as the fixed temperature detector. This type of fire detector will also operate on an abnormally rapid temperature rise over a short period of time. Frequently in the event of a fire the rate-of-rise actuation occurs before the fixed temperature is reached. Figure 7.4 shows a combination rate-of-rise/fixed temperature detector.

Smoke Detectors

Smoke detectors are another type of initiating device. They are actuated by products of combustion released in the very early stages of a fire. These products may be in the form of very tiny particles, invisible to the human eye, or they may be in the form of visible smoke.

There are two types of smoke detectors. *Ionization detectors* contain a small amount of radioactive material that ionizes the air passing through the detector. Products of combustion slow the ionization process, which causes the detector to go into alarm. There are very strict rules governing the installation (location, spacing, and area of coverage) of these detectors. Figure 7.5 shows an ionization smoke detector.

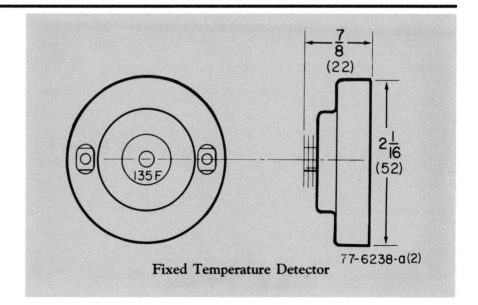

Fixed Temperature Detector

Figure 7.3

Photoelectric detectors, while similar in appearance to ionization sensors, operate only when visible smoke has been generated. A small beam of light is directed through a small, dark-colored cylinder. A photoelectric sensor is located in a branch off the cylinder. Smoke entering the cylinder will deflect the light to the photoelectric sensor, thereby causing the detector to go into alarm. See Figure 7.6.

Figures 7.5 and 7.6 depict ceiling-mounted smoke detectors. These detectors are primarily for open-area protection. Duct-mounted smoke detectors of both types are also available to sample air passing through air handling systems. These detectors can be used to close smoke dampers, shut down fans, and send an alarm signal to a central location. Figure 7.7 shows a duct-mounted smoke detector.

Water Flow Sensors

Sprinkler heads are similar to thermal detectors in that they actuate when they reach a fixed temperature setting. When a sprinkler head is actuated, water begins to flow in the piping system. A water flow switch, located in the sprinkler piping, is then actuated to cause an

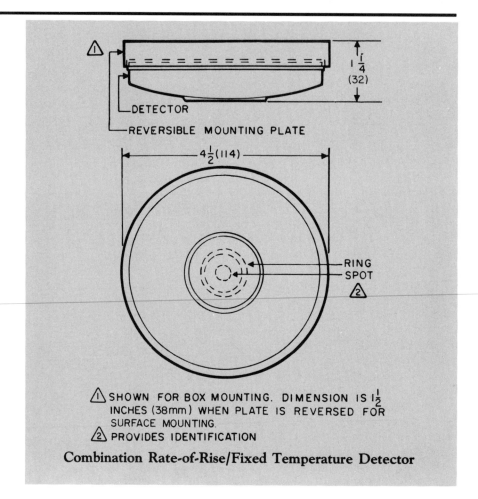

Combination Rate-of-Rise/Fixed Temperature Detector

Figure 7.4

alarm at a central location. Sprinkler heads, however, are not as sensitive as thermal detectors and therefore do not respond as rapidly as thermal detectors. This is because of the mass of the sprinkler heads and their associated piping. The advantage of this slightly slower reaction is that they do not activate so quickly, and thereby may avoid unnecessary water damage. Figure 7.8 shows two types of sprinkler heads.

Manual Pull Stations

Manual pull stations are not automatic fire alarm devices – they require a person to actuate them manually. These devices are normally

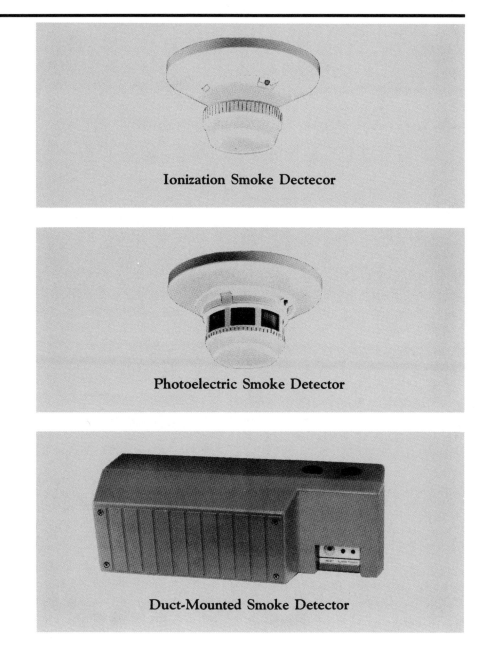

Figure 7.5

Ionization Smoke Dectecor

Figure 7.6

Photoelectric Smoke Detector

Figure 7.7

Duct-Mounted Smoke Detector

mounted in hallways and lobbies near exit doors. Figure 7.9 shows a manual pull station.

Indicating Circuit Devices
The indicating circuit in a fire alarm system contains devices that notify building occupants of the presence of fire. Bells and horns are commonly used to signal an alarm condition to building occupants. When bells are utilized for other purposes in a building, however, horns must be used for fire alarm notification. See Figures 7.10 and 7.11.

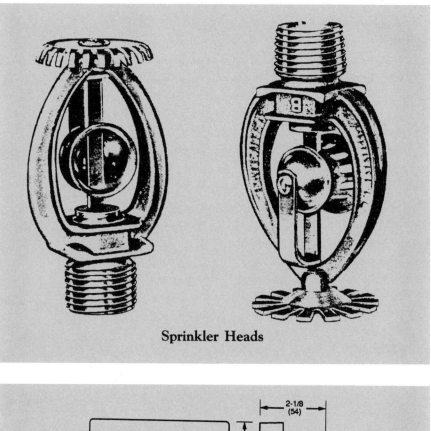

Sprinkler Heads

Figure 7.8

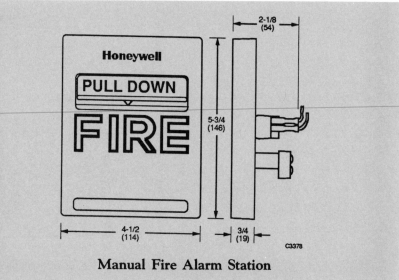

2-1/8
(54)

Honeywell

PULL DOWN

FIRE

5-3/4
(146)

4-1/2
(114)

3/4
(19)

C3378

Manual Fire Alarm Station

Figure 7.9

There are other types of signals used for alarm notification. In buildings where coded alarms are necessary, such as hospitals, chimes are used as indicating devices. Chimes also have a softer, less irritating sound than bells or horns. For outdoor applications, or indoor areas with loud background noise, sirens are commonly used. For the benefit of building occupants who have hearing difficulties, bells and horns are available with accompanying high-intensity strobe lights to provide visual indication of an alarm condition (See Figure 7.12). High-efficiency loudspeakers are also used as indicating devices in fire management systems. These speakers are capable of transmitting audible fire alarm signals and voice communication messages. They are suitable for indoor and outdoor installation.

Fire Alarm Control Units

Since a fire alarm system is in a standby or passive state for extensive periods of time, there should be constant indication that the system is operationally sound. Consequently, most of the control panel external circuits are electrically supervised on a continuous basis, to assure that they are in working order. If a circuit malfunction occurs, an audible trouble alarm will be energized, and a visual trouble alarm light will illuminate at the control panel. A fire alarm system that cannot operate due to a faulty circuit or device could result in destruction of property and/or loss of life. Therefore, the supervision of circuits is required to provide circuit continuity on a constant basis. Fire detection and alarm

Fire Alarm Horn

Figure 7.10

Fire Alarm Bell

Figure 7.11

panels are classified as *fire alarm control units* when their initiating circuits and indicating circuits are supervised. Fire alarm control units contain contacts for remote alarm and trouble indication. These contacts can be used for additional functions such as releasing fire doors, or remote station operation. Operation from stand-by rechargeable batteries with an automatic battery charger is also an option.

There are two types of fire alarm control units: *single-zone units* and *multiple-zone units*. A single-zone fire alarm control unit has one

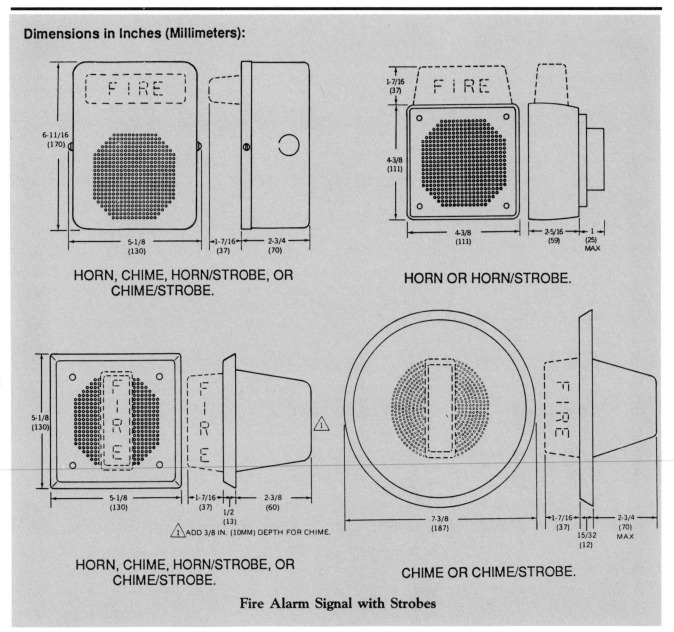

Fire Alarm Signal with Strobes

Figure 7.12

initiating circuit that is wired to multiple initiating devices, and one indicating circuit that is wired to multiple indicating devices. See Figure 7.13.

A *multiple-zone fire alarm control unit* consists of multiple, single-zone modules, as required, to meet the building's fire alarm zoning needs. Each module of the multiple-zone fire alarm control unit is autonomous and operates independently of the other zone modules. See Figure 7.14.

Fire management panels (Figure 7.15) consist of initiating circuits using the same single-zone modules as a multi-zone fire alarm control unit. In place of bells, horns, or chimes in the indicating circuits, fire management panels use speakers powered by amplifiers. An interface module provides command functions to the speaker zones. A telephone module with remote telephones allows firefighters to communicate between the control panel and remote locations. There is also an HVAC module that can control fans or dampers, as required, to contain smoke or fire in any zone.

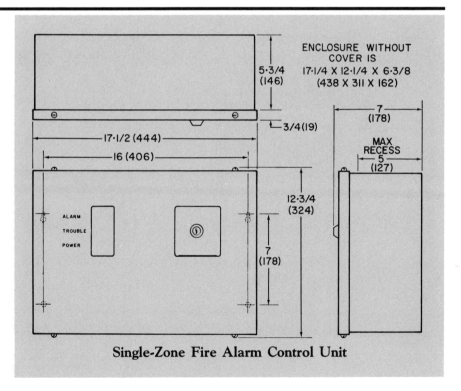

Single-Zone Fire Alarm Control Unit

Figure 7.13

Microprocessor-Based Fire Alarm System

The use of microprocessor-based distributed process system technology has led to the development of more sophisticated fire alarm systems. This technology has added intelligence to the fire alarm control unit to improve reliability and flexibility.

Stand-Alone Fire Alarm Control Unit

A basic stand-alone fire alarm control unit is a multiple-zone fire alarm control unit with its function modules mounted on plug-in circuit boards. The control unit monitors conventional fire sensors that are arranged by zones. All the variables of a multiple-zone fire alarm control unit are addressed by software programmed into the control board. The addition of an interface board, in most systems, allows a basic stand-alone fire alarm control unit to communicate, via a communication bus, to a higher level central host personal computer. However, if communication with the central host is lost, the stand-alone control unit will continue to perform its fire alarm initiating and indicating functions.

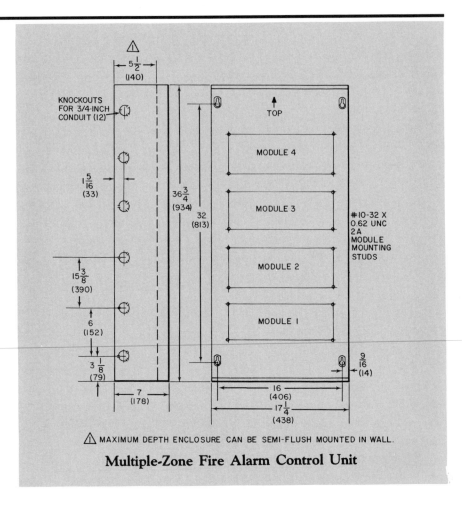

Multiple-Zone Fire Alarm Control Unit

Figure 7.14

Stand-Alone Fire Management System

This system consists of a basic stand-alone fire control unit, a fire command station, and a remote audio cabinet. The system's indicating circuits notify the building's occupants of a fire by broadcasting special fire tones or a voice message over speakers. Like the basic stand-alone fire alarm control unit, the fire management system can also communicate with a higher level central personal computer.

Stand-Alone Intelligent Fire Alarm System

This system uses intelligent initiating circuit sensors, such as thermal sensors or photoelectric and ionization smoke sensors. Intelligent indicating circuit devices are also used to provide software-driven fire alarm notification. Each intelligent initiating circuit sensor and indicating circuit device contains a custom integrated circuit, enabling two-way communication to a stand-alone intelligent fire alarm system control unit. Also, each sensor can communicate its individual point address (based on sensor type) and an analog value to the fire alarm control unit. In general, most systems can analyze the analog signal to measure each sensor's sensitivity and to determine its status, such as pre-alarm (signaling the need for maintenance), normal, trouble, or alarm state. Like the basic stand-alone fire alarm control unit, the intelligent fire alarm system can communicate with a higher level central host personal computer.

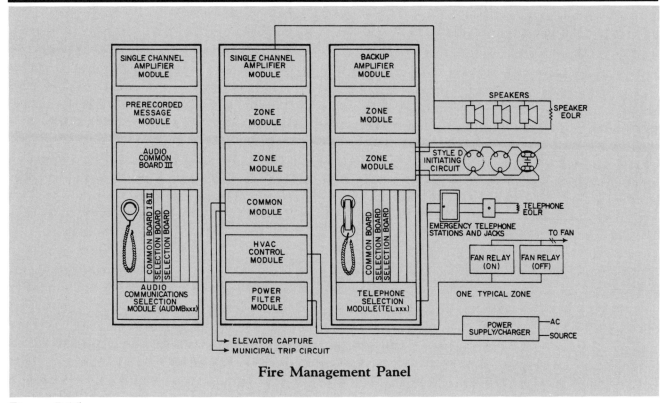

Fire Management Panel

Figure 7.15

Fire Alarm System Integration

A fire alarm system can be integrated with other building systems, such as heating, ventilating, and air-conditioning (HVAC), lighting, security, and access control, through a Building Automation System (BAS). The BAS can receive and process information from these various building systems. In the event of a fire, interactive software programs can activate sequences which will notify building occupants of a fire, control HVAC systems to contain smoke or provide smoke-free areas for occupants, take control of the elevators, and notify authorized personnel of the fire's location and what action to take.

There are two ways to incorporate a fire alarm system into a Building Automation System: 1) an existing, stand-alone, functioning fire alarm system can be connected to a BAS, or 2) a new state-of-the-art BAS-compatible fire alarm system can be installed.

Partial Integration

An existing fire alarm system is basically a stand-alone system that has an initiating circuit, an indicating circuit, and a control unit. It may also have a central operator interface, or monitor, to indicate fire location and monitor system status.

The control unit has auxiliary alarm contacts for each zone, and auxiliary contacts for system trouble indication. These auxiliary contacts are considered to be digital output points from the control unit. When hard-wired to a BAS system controller, they become digital input points. Each set of alarm or trouble contacts connected to a system controller is a digital input point. In the event of a fire, the control unit will actuate the indicating circuit, announce the fire's location and other pertinent information at its central monitor, and actuate the appropriate zone auxiliary switch. The BAS host computer will display and print preprogrammed information, such as time, date, fire location, and operator action messages. Through software programs, the BAS system controller can also activate HVAC control sequences such as smoke containment or smoke evacuation.

There are some advantages to integrating a fire alarm system through a BAS with other building systems. First, an existing fire alarm system, if functional, can be connected to the BAS at a considerable cost savings. Second, the interaction of software programs can add some life safety and property protection features to the existing fire alarm system. Third, it allows central monitoring of the fire alarm system and other building systems by one operator.

One disadvantage of partial integration is that the interconnecting wiring between the existing fire alarm control unit and the BAS controller is not supervised. As a result, there could be a failure of this wiring without notification. Figure 7.16 shows the connection of an existing fire alarm system to a Building Automation System.

Total Integration

In new building construction, or in renovation of existing buildings, serious consideration should be given to the total integration of as many building systems as possible. At that time it is possible to combine HVAC, lighting, fire, security, and access control under one central Building Automation System. Total integration of the fire alarm system means that a single central host computer serves the fire alarm system as well as the other building systems. The fire alarm control panels and the system controllers (HVAC and lighting) communicate with the central host computer over separate communications buses. The fire alarm control panels, like the HVAC system controllers, are designed to operate as stand-alone systems if the communications link with the central host computer is lost. The central host computer can monitor both the HVAC systems and the fire alarm systems. An optional fire alarm monitor can also be located remote from the central host to monitor only the fire alarm points. Figure 7.17 shows how a fire alarm system is integrated with HVAC systems.

One of the major benefits of total building systems integration is *systems interaction*. Assume, for example, that the fire alarm control units and HVAC system controllers are integrated through software. If a fire occurs on one floor of a multi-story building, the ventilating units can be used to prevent the smoke from spreading. Exhaust dampers can be opened and outdoor air intake dampers closed on the fire floor. On the floors above and below the fire floor, the exhaust dampers can be closed

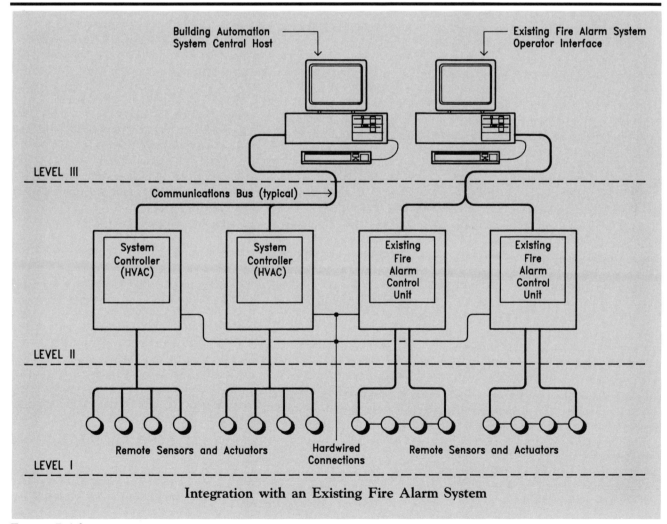

Integration with an Existing Fire Alarm System

Figure 7.16

and the outdoor intake dampers opened. This will pressurize the area surrounding the fire floor, thereby containing the smoke (See Figure 7.18). This is only one of many possible Fire/HVAC software strategies which can be employed through building systems integration.

Central Host Computer

The central host PC, when connected to intelligent fire alarm control panels, can monitor individual fire initiating devices, such as smoke detectors and thermal detectors. Consequently, alarm response and evaluation can be based on the initiating device location rather than multi-device zone location.

Color graphic software provides a library of standard symbols and graphics packages to simplify the on-line construction of graphics. Floor layouts can be created graphically that depict fire initiating devices in their actual locations. In the event an initiating device goes into alarm,

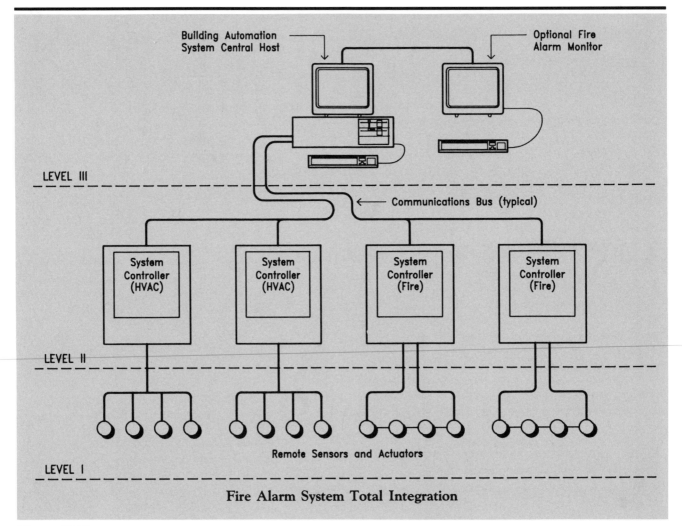

Fire Alarm System Total Integration

Figure 7.17

the appropriate floor layout can be displayed on the monitor (CRT), showing the operator the alarm location. The monitor can display emergency action instructions for the operator to follow, while the printer generates a hard copy of these instructions. The printers associated with the central system provide documentation of all pertinent system activity. In the event of a fire alarm, the printers record the type of alarm, the zone or individual sensor location, the time of the alarm, an operator acknowledgment code, and the time of acknowledgment.

Event-initiated programs can activate interactive software programs among integrated building systems. A fire alarm system can interact with HVAC systems to prevent the migration of smoke to other zones or floors. A fire alarm can also interact with the access control system to unlock secured doors and allow building occupants a means of egress.

Alarms can be prioritized, enabling more critical alarms to supersede less critical alarms. For example, if an HVAC fan failure alarm occurs simultaneously with a fire alarm, the fire alarm will appear first. When the fire alarm has been acknowledged, the fan failure alarm will display. Fire alarms from critical areas can be programmed to override fire alarms from areas or zones not considered to be as critical.

Most BAS's have multiple levels of operator access to the system through the central host terminal. These authorization levels are assigned by the system owner. Operators sign on with an ID and

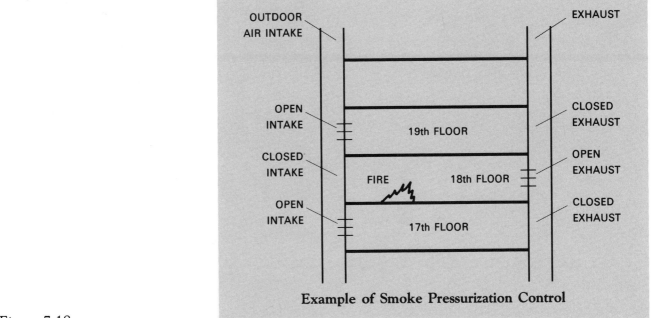

Example of Smoke Pressurization Control

Figure 7.18

password, which define that operator's level of access. For example, an operator may only be authorized to receive and acknowledge alarms, not to make program changes. This feature prevents operators from intentionally or inadvertently deleting or changing fire alarm programs.

Integrating a building fire alarm system with other building systems can offer many benefits. Interactive software programs can increase the quality of life safety programs, and consolidating building monitors and alarm activities at a central location offers some economic and operational benefits as well. There may also be some advantages in the installation phase from fire alarm system integration.

Chapter Eight

Security-Access Control Integration

The theft, vandalism, and violence which permeate our society today have increased the demand for security systems and services. There was a time when simply locking the door at the end of the day was sufficient protection. Today, the worldwide security market is a multi-billion-dollar industry that continues to grow annually at a tremendous rate.

Businesses have special security needs. Inventory in warehouses and stockrooms must be protected against theft from both internal and external sources. Confidential records and documents must be protected from unauthorized access. Computer rooms, which are critical to the operation of most companies today, make tight security and personnel access control extremely important. The threat of vandalism alone makes security around the perimeter of any building almost mandatory. Some businesses have building zones that must be secure on a 24-hour basis, while in other buildings, the entire system can be deactivated during occupied periods and activated during unoccupied periods.

Building Security Systems

Major advances in microelectronics have resulted in the development of more sophisticated building security systems. Not long ago, magnetic switches on perimeter doors and windows comprised a building's entire security system. Today, this basic form of intrusion protection represents only a portion of a comprehensive building security system. Extremely sensitive motion detectors located in interior zones of a building can detect even the slightest movement of a person or object in the protected area. Pulsed, infrared, and photoelectric beam sensors have been developed to replace the old lamp-type sensors that were easy to compromise. Many other types of security sensors, such as pressure mats, glass break detectors, vibration detectors, taut wire detectors, and audio detectors, are now available to effectively render a building secure.

Categories of Protection

Building security systems provide three categories of protection, which may be thought of as lines of defense. These lines of defense are: 1) perimeter protection, 2) area protection, and 3) object protection.

Perimeter protection is protection against unauthorized intrusion into a building through exterior walls, windows, doors, transoms, skylights, or ventilating ducts. Detection of this type of entry can be sensed by vibration detectors, magnetic door and window switches, and glass door and window tape.

If perimeter penetration is achieved, a second line of defense is *area protection*. The primary goal of this type of protection is to detect movement in the protected area. Motion detectors, infrared detectors, and photoelectric beams can be used for movement detection.

A third line of defense is *object protection*. This method of protection is used where a specific item or object is to be protected. Filing cabinets, safes, or art objects fall into this category. Proximity sensors, infrared detectors, and vibration detectors are commonly used as sensors for this type of protection.

Basic Security System Components

All security systems must be capable of sensing unauthorized intrusion and providing notification to some responsible person or persons. There are also a number of functions, directly or indirectly related to security, that the security system can perform, such as elevator alarm, panic alarm, or hold-up alarm, releasing doors to allow people to leave the building or securing them in a protected area.

All building security systems contain basic components. A basic security system consists of a control unit, initiating devices (a circuit to sense unauthorized intrusion), and indicating devices (a circuit to provide an indication of the alarm). See Figure 8.1. The relay shown in Figure 8.2 is located in the control unit, the door and window switches in the initiating circuit, and the horn in the indicating circuit. The batteries are the power source. If a door or window is opened, the respective door or window switch will open, interrupting power to the relay in the control unit. The relay contacts will go to their normally closed position, connecting the alarm battery to the horn, thereby causing an alarm.

Security Initiating Circuit Devices

Initiating circuits operate with switching contacts that either close or open under alarm conditions. This may be as simple as a magnetic switch that opens or closes when a door is operated, or as complex as an electronic alarm panel connected to numerous sensing devices. Regardless of how it is accomplished, the initiating circuit is either opened or short-circuited in the event of an alarm condition.

Initiating circuit devices fall into several categories.

Contact Devices
- Magnetic door and window switches
- Mechanical switches (roller, butt, etc.)
- Window tape
- Duress switches
- Pressure mats
- Glass break detectors
- Vibration detectors
- Thermal switches
- Taut wire detectors
- Traps

Audio Detectors
- Ultrasonic sensors
- Audible sound devices
- Electronic vibration detectors

Microwave Detectors

Infrared Detectors
- Passive infrared
- Photoelectric beam

Security Indicating Circuit Devices

Security systems in large buildings do not use the same type of local indicating devices or alarms as those used in residential security systems. In some commercial buildings, a monitor or annunciator at a central guard station is considered to be the indicating device and control unit.

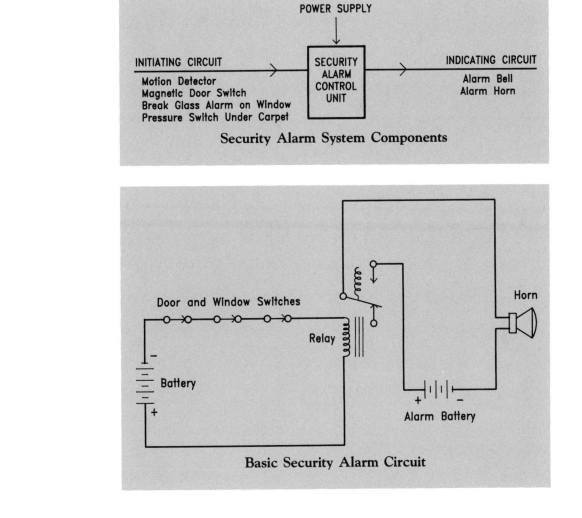

Figure 8.1

Security Alarm System Components

Figure 8.2

Basic Security Alarm Circuit

Security Control Unit

In buildings with stand-alone security systems, the initiating circuit devices communicate directly to a central monitoring station (security control unit). The central monitoring station acts as the security control unit and indicating device.

Access Control Systems

In the past, access to a building or area was controlled by security guards, locked doors, and fences. The advent of closed circuit television (CCTV) made it possible for one centrally located person to guard a number of doors or gates through electrically controlled locks.

Card reader access control systems are an effective means of controlling personnel movement in a building. The cards are issued to select individuals and read by a card reader which, in turn, unlocks doors assigned to the card holder. These systems limit access to restricted areas to authorized personnel only. After-hours access to a building can be restricted to specific cardholders and specific doors. The card reader access control system can also be used for time and attendance record keeping. The concept of using coded cards has been available since the late 1950s. Newer systems have been developed to provide a much higher security level. For example, recent technology has made possible the use of a person's physical features, such as a handprint, voice, or eye characteristics, to identify the person and allow access.

Figure 8.3 illustrates a typical access control system.

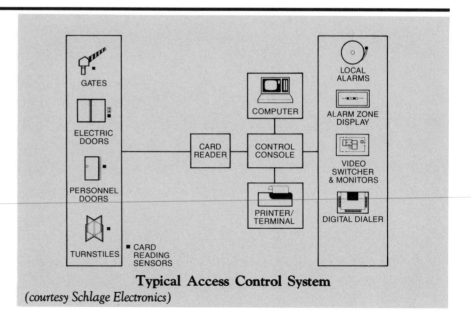

Typical Access Control System
(courtesy Schlage Electronics)

Figure 8.3

Building Automation System (BAS) Integration

Most of the security and access control systems on the market today can operate in a stand-alone manner. However, the trend in new buildings is toward the total integration of building systems. To some degree this integration exists today. Through total system integration, a BAS can receive information from a security access control system. This is shown in Figure 8.4.

The integration of security and access control, and other building systems (such as HVAC, lighting, fire) into a Building Automation System can provide both economic and operational benefits. First, initial installation work, such as electric wiring, can be consolidated, resulting in cost savings. Substantial paybacks can be generated through HVAC energy management and lighting programs, thereby offsetting some of the costs involved in the integration process. Second, the cost

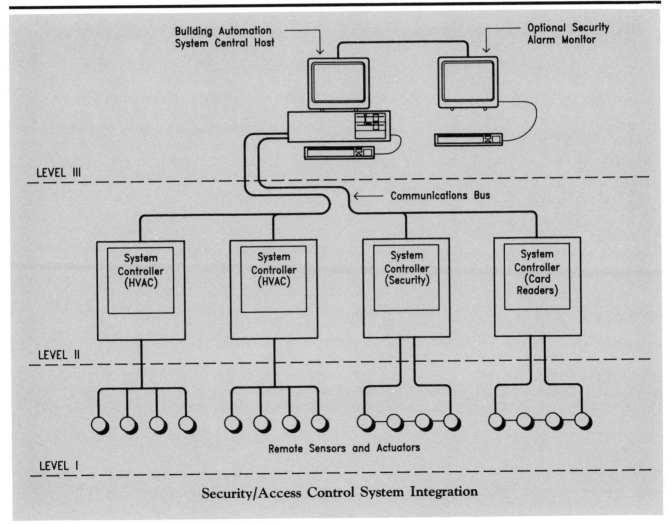

Figure 8.4

of on-site guard services can be greatly reduced. After-hours monitoring of all alarms (HVAC, fire, security), can be accomplished by one operator from a single, central location. Third, interactive software programs between security-access and fire alarm systems can, in case of fire, unlock perimeter doors, providing a means of emergency egress from the building. Finally, independent terminals and printers, serving each system, can be provided at different locations.

Central Host Computer

A single central host personal computer serves the security-access system, fire, and HVAC system controllers. The Level II security-access control units, fire, and HVAC system controllers communicate over separate communication buses. The Level II controllers are designed to operate as stand-alone systems if the communication link with the central host PC is lost. The Level II security-access control units can also provide annunciation locally, or to a remote or central station system. The central host terminal and printers can be located at a security office and, if desired, limited to security-access information. However, the security-access control system, fire, and the HVAC system controllers are interactive through software. Upon activation of a security-access initiating sensor, the central host can provide the following:

1. A description of the zone or initiating sensor detecting an unauthorized intrusion appears on the CRT and printer. The CRT may also show an area, building, or floor layout indicating the location of the intrusion alarm. Emergency action instructions are also displayed on the CRT.
2. A hard copy is printed out documenting the location of the unauthorized intrusion, time and date, and alarm acknowledgment by whom and when.
3. Operator can override exit doors, locked by the security-access system, to permit entrance or egress during emergency or unusual situations.
4. CRT displays (with a hard copy printout) a list of cardholders entering and leaving a secured area by card number, time, date, and location.

The integration of security and access control systems with other building systems offers an opportunity for cost savings on the initial installation and ongoing operating costs. The consolidation and monitoring of all building systems at a central operating center provides some obvious economic benefits. The interactive software program potential of total integration creates an enhanced quality of life safety and property protection. .

Chapter Nine

Facility Management Programs

A Building Automation System (BAS) is a tool that can be used to assist facility management and operating personnel in four important ways. First, Building Automation Systems maintain comfort conditions in a building. BAS are designed to control HVAC systems in the most efficient manner, while providing adequate ventilation and comfortable temperature and relative humidity levels. Microelectronic sensors feed signals to microprocessor-based controllers, which control environmental conditions to within very close tolerance levels. As a result, building occupants are exposed to only the highest level of indoor air quality.

Second, Building Automation Systems control operating costs. For example, BAS energy management software programs control energy-consuming systems (HVAC and lighting) in the most cost-effective manner. These energy management programs are included in the standard software offered by most manufacturers today. Significant energy dollars can be saved without sacrificing comfort conditions by activating these energy-saving strategies.

Third, Building Automation Systems monitor other building systems. BAS is designed for total integration of HVAC systems, lighting, fire safety, and security and access control. The advantages of integrating these building systems are: lower initial cost, lower operating cost, improved life safety, reduced control system maintenance requirements, and the ability to monitor and manage all building systems from a central location.

Finally, a building automation system has the ability to generate management reports from information supplied to it from remote sensors and input devices. These management reports are a useful tool for extending equipment life expectancy, controlling operating costs, scheduling preventive maintenance tasks, controlling inventory, monitoring efficiency of energy-consuming systems, and recording energy consumption for an individual zone or for the entire building. This feature is discussed in more detail in the next sections.

Computerized Maintenance Management Programs

In general, many third-party computerized maintenance management software programs now available schedule maintenance activity on a real-time basis. The monitoring and reporting features provide facility management personnel with the tools needed to protect equipment, control costs, schedule workloads, review historical trends, manage materials, and plan budgets. The program issues work orders either automatically or by operator request, based on calendar time, equipment run time, or event occurrence. CRT display and hard copy printouts depict detailed work status, delinquent maintenance, inventory, and financial statistics reports. Supervisors can assign system access levels so that maintenance personnel will have access to information that pertains to their job classification only.

Maintenance Scheduling

The operators can set up *tasks* based on user-defined material, skill, and tool data bases. Multiple tasks are linked together for a *maintenance activity* on a specific piece of equipment. Maintenance schedules for a maintenance activity, or *job* depict detailed work orders for each associated task. Figure 9.1 shows the maintenance activity/job process.

Figure 9.2 is a work order that can be scheduled on a calendar basis (every six months), equipment run time basis (every 2,000 hours of run time), or event-occurrence basis (dirty filter alarm). An operator can also request a work order manually.

Work Order Printout

Maintenance schedules are printed out automatically at a user-defined interval: daily, weekly, monthly, or yearly. The work orders can be sent to a remote printer or located in the department responsible for the work.

In general, maintenance requests consist of *task control* and *work order pages* (Figure 9.3). The task control and work order pages contain all pertinent information relating to the work to be performed. Overview information includes a work order reference number, issue date, supervisor, requesting person, due date, last work order reference number, and other descriptions. When all work is completed, the maintenance person fills out the "Comments" area of the work order, noting the materials expended and time required to complete the work.

Next, the completed work order is processed by the system operator. If the maintenance is delayed, the operator can list it in backlog or, if the work is not required, cancel the work order. Figure 9.4 illustrates the work order mechanism (from issue to completion of the work order).

Maintenance History

Most maintenance management software programs can archive all work order data. The operator can then retrieve historical information about equipment, skills, materials, and tasks.

Material Inventory

A maintenance management software program may include a record keeping system for maintenance materials. Figure 9.5 depicts an inventory control printout which is used to track material quantity, costs, usage, and stock location. When material inventories drop below a predetermined level, the program issues a reorder request, as shown in Figure 9.6. This ensures that the parts and materials are available to maintenance personnel on an as-needed basis.

Financial Analysis

A maintenance management software program allows the system operator to set up monthly and annual budgets for any or all system requirements such as skills, materials, equipment, tasks, maintenance activities, and jobs. (See Figures 9.7 and 9.8.) Projected budget information can be compared with accumulated actual costs to show any deviation between actual vs. planned system performance. As a work order is completed, the software automatically allocates maintenance expenses and assigns costs to the associated system or system component.

Management Information

Most maintenance management software programs include CRT displays and hard copy, standard reports depicting summaries of maintenance tasks completed, as well as work order status.

Maintenance summaries include costs and usage for like component materials, skills, tasks, and maintenance activities. Work order status includes information on in-progress, backlogged, delinquent, cancelled, and completed maintenance projects.

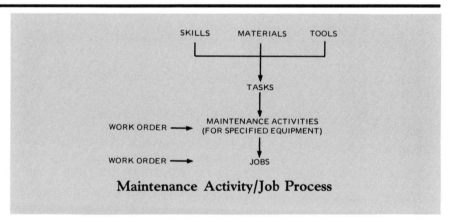

Maintenance Activity/Job Process

Figure 9.1

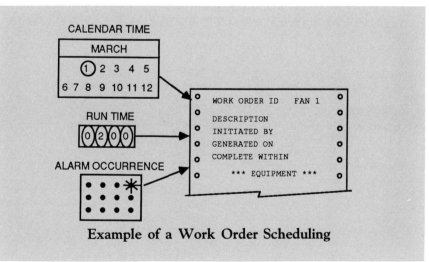

Example of a Work Order Scheduling

Figure 9.2

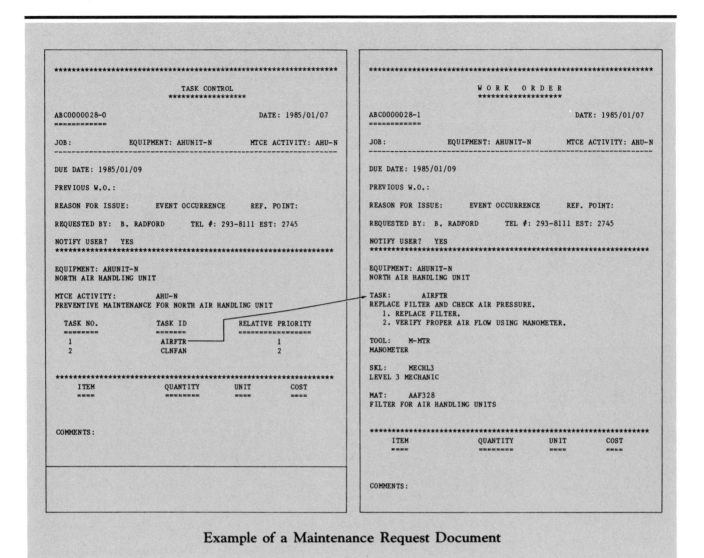

Example of a Maintenance Request Document

Figure 9.3

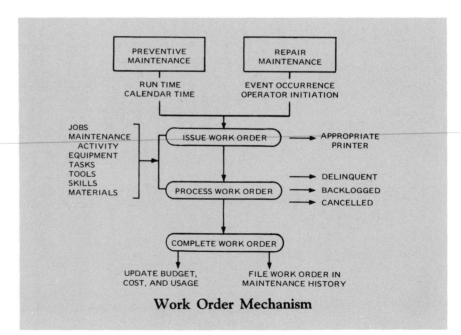

Work Order Mechanism

Figure 9.4

Utilities Metering Program

It is frequently the responsibility of the building manager, director of facilities, or a member of the building owner's staff to periodically review and keep ongoing records of energy consumed by the facility. These records include the amount of electricity or fuel (natural gas or oil), purchased steam, or any other heating and/or cooling medium purchased from an outside source. In many cases, the building owner or his operating staff must rely on the accuracy of the vendor's meter for their energy consumption records. However, many times a single meter records energy consumption for an entire facility with multiple tenants. The question then is how to fairly allocate costs to each tenant. Energy costs are sometimes allocated on a square foot basis or on the basis of occupancy hours. Unfortunately, there is no accurate way of distributing the costs equitably unless individual meters are installed for each tenant.

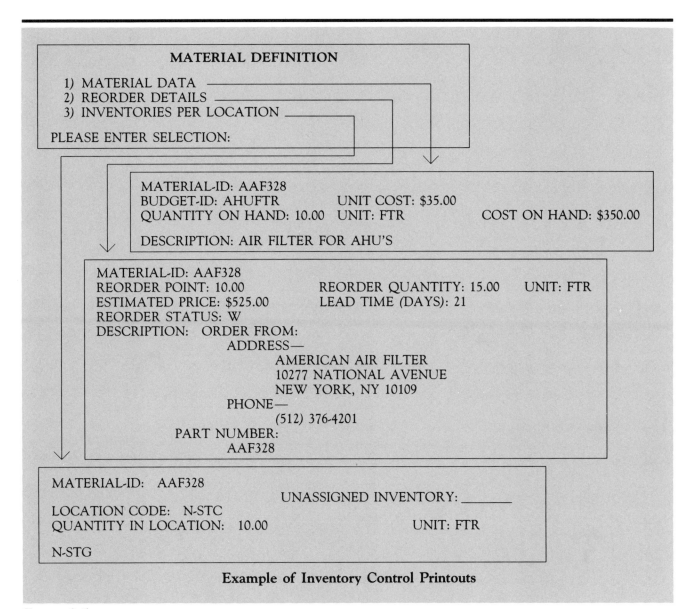

Example of Inventory Control Printouts

Figure 9.5

147

MATERIALS TO REORDER

Material ID	Unit	Quantity On Hand	Cost On Hand	Quantity To Reorder	Estimated Price
MAT01	EA	100	$ 200	200	$ 2
MAT04	EA	2	$ 250	10	$ 150
MAT05	EA	100	$ 200	200	$ 2
MAT06	EA	5	$ 170	30	$ 34
MAT08	DOZ	12	$ 4200	20	$ 350

*** END OF REPORT ***

MATERIALS ON ORDER

1985/05/20

Material ID	Unit	Quantity On Hand	Cost On Hand	Quantity Ordered	Estimated Price
MAT02	EA	100.00	$ 200.00	200.00	$ 2.00
MAT07	EA	5000.00	$ 50000.00	50.00	$ 10.00
MAT10	GAL	13.50	$ 1188.00	25.00	$ 88.50

*** END OF REPORT ***

Example of Material Reordering Reports

Figure 9.6

Budget ID: AHUFTR

Budget Year: 1991
Unit: FTR

	Costs		Maint. Schedules		Quantities	
	Plan	Actual	Plan	Actual	Plan	Actual
Jan	$ 35.00	$ 70.00	01	02	01.00	02.00
Feb	$ 00.00	$ 00.00	00	00	00.00	00.00
Mar	$ 35.00	$ 35.00	01	01	01.00	01.00
Apr	$ 00.00	$_____	00	___	00.00	_____
Jun	$ 35.00	$_____	01	___	01.00	_____
Jul	$ 00.00	$_____	00	___	00.00	_____
Aug	$ 35.00	$_____	01	___	01.00	_____
Sep	$ 00.00	$_____	00	___	00.00	_____
Oct	$ 35.00	$_____	01	___	01.00	_____
Nov	$ 00.00	$_____	00	___	00.00	_____
Dec	$ 35.00	$_____	01	___	01.00	_____

Example of a Monthly Budget Data Report

Figure 9.7

A Building Automation System provides the means to dynamically monitor and record a facility's energy consumption on a real-time basis. The strategic installation of utility meters by individual tenant, department, floor, zone, or mechanical system provides a method of fairly assessing a tenant's energy consumption. In addition, the BAS can monitor the outdoor air as well as space temperature and relative humidity. This information, combined with the building energy consumption data, can be used in historical trend graphs to determine the impact on energy consumption by these measured variables. The wealth of building operating cost information that can be provided by a BAS can help a building owner or operating staff to accurately define energy consumption areas, and make more informed decisions on how to control the energy consumption of HVAC and lighting systems.

Tenant Energy Monitoring Program

Air-conditioning in most commercial buildings today is provided by constant volume single zone air handling systems, heat pumps, or variable air volume systems. Most of the automatic temperature control companies manufacture direct digital control (DDC) controllers which regulate zone level terminal units supplied by these systems. These terminal units may be reheat coils, variable air volume boxes, or zone heat pumps. The DDC zone controllers, linked by a communications bus, can send air flow and temperature information to a central host personal computer. The computer, through energy algorithms, can calculate heating or cooling energy supplied to a tenant zone or floor. This provides an excellent means of determining how much energy a tenant consumed during overtime occupancy periods. The BAS can then automatically invoice tenants on a monthly basis for energy consumed in their areas. The invoice will document the invoice date, time period, and energy consumption.

Figure 9.9 shows how a variable air volume box can be monitored for energy consumption. The temperature of the air entering the box (called *primary air*), space temperature, and air flow through the box are monitored by the DDC zone controller over a specified period of time. The central host computer, using these variables, computes the energy consumed per hour and applies the number of hours of operation to determine the total amount of energy consumed during that time period. The computer then applies to this figure the cost per energy unit to determine total cost of energy used in that time period.

Budget ID: AHUFTR	Year: 1991	Division: 870	Dept: 394	Acct: 799
	—Planned—		—Actual—	
Costs	$ 210.00		$ 105.00	
Maint. Schedules	6		3	
Quantities	1		1.5	

Description: Filter for Air Handling Units

Example of a Yearly Budget Data Report

Figure 9.8

Heating/Cooling Plant Efficiency Program

One of the primary goals of a building or facilities manager is to provide a comfortable, clean, and safe environment in the buildings he or she manages, and to do so in the most cost-effective manner. The rising cost of energy, and the changes to ventilation codes to improve indoor air quality, make this task extremely difficult. As previously stated, a sound preventive maintenance program on all heating and cooling equipment is the first step toward achieving this goal.

The next logical step is to continuously monitor the efficiency of the central heating and cooling plants (boilers and chillers). These primary heating and cooling sources frequently represent the largest portion of the energy bills for the physical plant. A small decrease in operating efficiency of these large central systems can result in a significant increase in energy consumption and its associated costs. Therefore, it is important that plant operating personnel be aware of any downward trend in plant operating efficiency so that corrective action can be taken immediately.

The same sensors used to monitor and control energy consumption can be used to monitor the efficiency of the heating or cooling plant. The simple equation for efficiency is:

$$\% \text{ Efficiency } = \frac{\text{output}}{\text{input}}$$

These sensors are strategically located in a central heating or cooling plant. Through a DDC controller, they constantly transmit critical

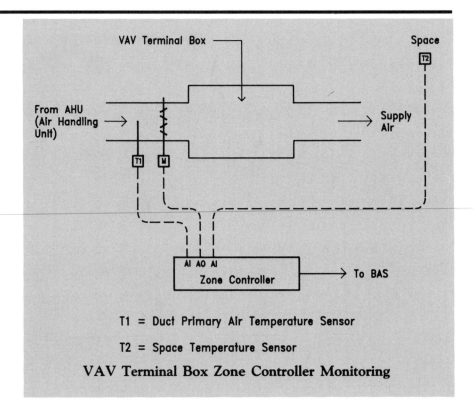

T1 = Duct Primary Air Temperature Sensor

T2 = Space Temperature Sensor

VAV Terminal Box Zone Controller Monitoring

Figure 9.9

150

information to a central host computer. This information can be used by the computer to develop energy consumption spreadsheets, graphs, or charts. The building manager or owner can then use this valuable information to monitor energy consumption trends and plan budgets.

Fossil Fuel Steam Heating Plant
Figure 9.10 depicts a method of monitoring the efficiency of an oil/gas-fired steam heating plant. Oil or gas input to the dual burner is measured to complete the input side of the equation. A system flow meter measures output from the boiler. Figure 9.11 illustrates a method used by the BAS to calculate percent efficiency.

Fossil Fuel Hot Water Heating Plant
Monitoring the efficiency of a hot water boiler is similar to that shown for the fossil fuel steam heating plant. The flow of oil and natural gas is measured to provide the input side of the equation. The boiler output, however, is determined by measuring water flow through the boiler and the temperature pickup of the water as it passes through the boiler. Figure 9.12 shows a method of monitoring the efficiency of an oil/gas-fired hot water heating plant. Figure 9.13 illustrates a method used by the BAS to calculate percent efficiency.

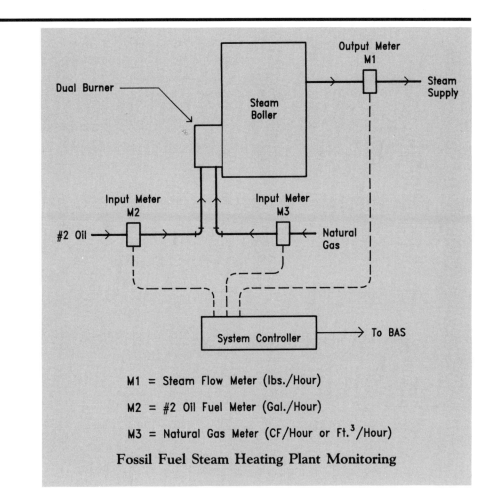

M1 = Steam Flow Meter (lbs./Hour)

M2 = #2 Oil Fuel Meter (Gal./Hour)

M3 = Natural Gas Meter (CF/Hour or Ft.³/Hour)

Fossil Fuel Steam Heating Plant Monitoring

Figure 9.10

Step 1:   Determine Consumed Input Energy (BTU's)

#2 Oil:

$$\underbrace{\frac{(Gal)}{(Hour)}}_{\substack{\text{Meter} \\ \text{M2} \\ \text{Fig. 9.10}}} \times \underbrace{\frac{139,600\ (BTU)}{(Gal)}}_{\substack{\text{Conversion} \\ \text{Factor}}} \times \underbrace{(Hours)}_{\substack{\text{User-defined} \\ \text{Time} \\ \text{Period}}} = \#2\ Oil\text{-}BTU's$$

Meter
M3

Natural Gas:

$$\overbrace{\frac{(ft^3)}{(Hour)}} \times \overbrace{\frac{950\ to\ 1,150(BTU)}{(ft^3)}} \times \overbrace{(Hours)} = Nat.\ Gas\text{-}BTU's$$

Step 2:   Determine Used Output Energy (BTU's)

Steam:

$$\underbrace{\frac{(\#)}{(Hour)}}_{\substack{\text{Meter} \\ \text{M1} \\ \text{Fig. 9.10}}} \times \underbrace{\frac{1,000\ (BTU)}{(\#)}}_{\substack{\text{Conversion} \\ \text{Factor}}} \times \underbrace{(Hours)}_{\substack{\text{User-defined} \\ \text{Time} \\ \text{Period}}} = Steam\text{-}BTU's$$

Step 3:   Determine Percent Efficiency (% Eff.)

$$\%\ Eff. = \frac{Output}{Input} \times 100\% = \frac{\substack{\#2\ Oil \\ or \\ Nat.\ Gas\ (BTU's)}}{Steam\ (BTU's)} \times 100\%$$

Fossil Fuel Steam Heating Plant
Percent Efficiency Calculation

Figure 9.11

Electric Chiller Plant

Monitoring the efficiency of an electrically-driven chiller plant is similar to that shown for the fossil fuel hot water plant. Electrical power consumed by the chiller is measured for the input part of the equation. The output is determined by measuring water flow through the evaporator, and the temperature difference of the water as it passes through the evaporator. Figure 9.14 shows a method of monitoring the efficiency of an electrically driven chiller. Figure 9.15 shows a method used by the BAS to calculate percent efficiency.

The engineering units for the variables are clearly different. Steam flow is measured in pounds per hour, while fuel oil is measured in gallons per hour, water flow in gallons per minute, electrical consumption in kilowatt hours, and natural gas in cubic feet per hour. Conversion of these units so that they can be used in the efficiency equation takes place in the central host computer software program.

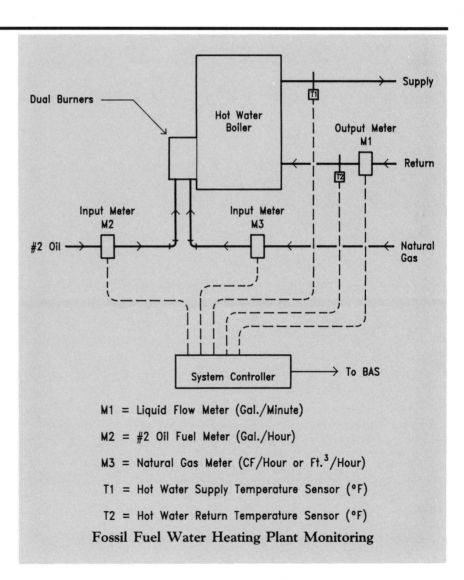

M1 = Liquid Flow Meter (Gal./Minute)

M2 = #2 Oil Fuel Meter (Gal./Hour)

M3 = Natural Gas Meter (CF/Hour or Ft.³/Hour)

T1 = Hot Water Supply Temperature Sensor (°F)

T2 = Hot Water Return Temperature Sensor (°F)

Fossil Fuel Water Heating Plant Monitoring

Figure 9.12

153

Step 1:   Determine Consumed Input Energy (BTU's)

#2 Oil: $\dfrac{(Gal)}{(Hour)} \times \dfrac{139,600\ (BTU)}{(Gal)} \times (Hours) = \#\ 2\ Oil\text{-}BTU's$

                    Meter
                    M2

               Fig. 9.12   Conversion   User-defined
                             Factor   Time
                                          Period

               Meter
               M3

Natural Gas: $\dfrac{(ft^3)}{(Hour)} \times \dfrac{950\ to\ 1,150\ (BTU)}{(ft^3)} \times (Hours) = Nat.\ Gas\text{-}BTU's$

Step 2:   Determine Used Output Energy (BTU's)

Hot Water: $500 \times \dfrac{(Gal)}{(Min)} \times (T1 - T2)(°F) = \dfrac{(BTU)}{(Hour)}$

                     Constant   Meter   Fig. 9.15
                                  M1

                         Fig. 9.12

$\dfrac{(BTU)}{(Hour)} \times (Hours) = Hot\ Water\ BTU's$

                    User-defined
                    Time Period

Step 3:   Determine Percent Efficiency (% Eff.)

$$\%\ Eff. = \dfrac{Output}{Input} \times 100\% = \dfrac{\#\ 2\ Oil\ or\ Nat.\ Gas\ (BTU's)}{Hot\ Water\ (BTU's)} \times 100\%$$

Fossil Fuel Hot Water Heating Plant Percent Efficiency Calculation

Figure 9.13

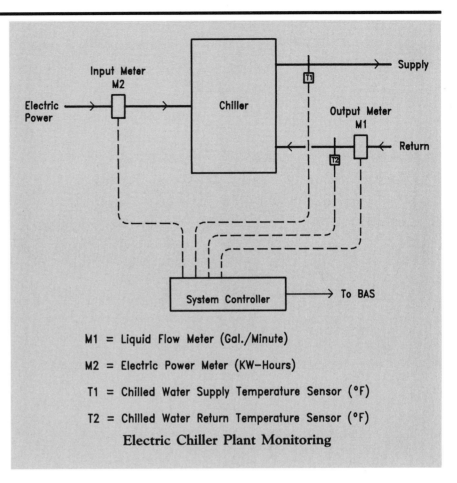

M1 = Liquid Flow Meter (Gal./Minute)

M2 = Electric Power Meter (KW-Hours)

T1 = Chilled Water Supply Temperature Sensor (°F)

T2 = Chilled Water Return Temperature Sensor (°F)

Electric Chiller Plant Monitoring

Figure 9.14

Step 1: Determine Consumed Input Energy (BTU's)

Electric Power: $(kw) \times \dfrac{(Ton)}{.77 \text{ to } .93 \text{ (kw)}} \times \dfrac{12{,}000 \text{ (BTU)}}{(Ton)(Hour)} \times (Hour) = \text{Electric-BTU's}$

$\underbrace{}$ Meter
M2
Fig. 9.14

$\underbrace{}$ Conversion Factors

$\underbrace{}$ User-defined
Time
Period

Step 2: Determine Used Output Energy (BTU's)

Chilled Water: $500 \times \dfrac{(Gal)}{(Min)} \times (T2 - T1)\,(°F) = \dfrac{(BTU)}{(Hour)}$

$\underbrace{}$ Constant

$\underbrace{}$ Meter
M1
Fig. 9.14

$\underbrace{}$ Fig. 9.14

$\dfrac{(BTU)}{(Hour)} \times (Hours) = \text{Chilled Water-BTU's}$

$\underbrace{}$ User-defined
Time
Period

Step. 3: Determine Percent Efficiency (% Eff.)

$\% \text{ Eff.} = \dfrac{Output}{Input} \times 100\% = \dfrac{\text{Electric-BTU's}}{\text{Chilled Water-BTU's}} \times 100\%$

Electric Chiller Plant Percent Efficiency Calculation

Figure 9.15

Part Three

Planning and Estimating for BAS

The demand for improved environmental conditions in buildings has escalated over the past 15 years. This, combined with increased utility costs, competition in the office space rental market, the emphasis on indoor air quality, and the demands of building owners for better methods of control and management of building systems, has compelled the manufacturers of HVAC equipment to reevaluate their products. The automatic temperature control industry responded to these needs with some major innovations.

The development of microelectronics and advancements in digital technology revolutionized building environmental control systems. Miniaturization, reduced cost, flexibility, accuracy of control, and advanced software programs made this technology ideal for BAS.

To determine whether or not your building is receiving the maximum benefits from its current control system, take a closer look. The existing pneumatic control system may appear to be functioning properly. However, slow response, inadequate maintenance, calibration shifts, and mechanical wear, although not obvious, can result in wasted energy and poor comfort conditions.

The Direct Digital Control (DDC) systems on the market today offer accurate control, the ability to perform complex control sequences, low maintenance requirements, and flexibility — some impressive features unachievable with traditional pneumatic controls. DDC systems can also perform energy management strategies and, if connected to a central host personal computer, provide total integration of other building systems. For new buildings, DDC offers the latest in automatic control technology. DDC controllers are available for use with all types and sizes of systems, from simple heat pumps to large built-up systems. The standard software in these controllers can perform basic control loops as well as the most complex control sequences. They can also perform all the traditional energy management strategies.

DDC controllers can operate as stand-alone units, performing all local HVAC control functions, with the option of adding a central host personal computer at a future time. If management reporting, HVAC system monitoring, or integration of other building systems (lighting, fire, security, and access control) are required, a personal computer can be included in the initial installation.

DDC sensors are more sensitive than pneumatic sensors and therefore respond faster and more accurately. DDC devices in general are not mechanical; they have no moving parts and therefore require minimum maintenance. Calibration of DDC sensors is usually not necessary.

The ideal time to convert to a DDC system in an existing building is when the HVAC system is being upgraded. When a control system is replaced, it is usually not necessary to replace all existing control devices. Valve and damper actuators, low temperature controllers, smoke and fire detectors, and many other devices, if functional, can be made compatible with the DDC controller. Chapter 10 examines the logical method of constructing a DDC system with a central host computer. It also reviews budget estimating using *Means Mechanical Cost Data*.

Chapter Ten

System Layout and Budget Estimate

This chapter presents an overall review of HVAC system layout, and the sequences of operation. The final section of this chapter shows how to prepare a budget estimate using a portion of the BAS system as an example.

System Layout

DDC controllers contain all of the software required to control HVAC systems and perform energy management strategies. A microprocessor replaces the pneumatic or electric analog controller, which connects the sensor with the actuator. The controller can receive input signals in one of two forms – digital or analog – and can take action through digital and analog output signals. These input and output signals are commonly referred to as *points*.

A digital input (DI) point is a switching action input (open-closed or on-off). An example would be a duct-mounted static pressure switch, with switching contacts that close when there is an increase in static pressure and open when static pressure drops below its setting. This device is commonly used to signal fan status (on-off) to a controller.

An analog input (AI) point is a variable input signal sent to a controller from a sensor that measures flow, temperature, pressure, or relative humidity. The input signal may be 500 ohms, 2 to 10 volts DC (VDC), or 4 to 20 milliamps (MA).

A digital output (DO) point is a two-position switching action by the controller. This switching action can be used to start and stop a fan or pump. It can also control a damper or valve actuator in a two-position (open-closed) manner.

An analog output (AO) point is a variable, or modulating output signal, from the controller that can modulate a damper or valve actuator. It can also be used to modulate a variable fan speed controller. Output signals can be 3 to 15 psi, 2 to 10 volts DC (VDC), or 4 to 20 milliamps (MA).

Software programs are routines that make the microprocessor in a DDC controller perform specific control functions. Input signals are converted to digital form and stored for future transmission. Or, the signals are converted from digital form to analog or digital output signals.

Standard software programs for HVAC applications are resident in DDC controllers. These programs can be configured and modified to fit a specific application. If a standard software program is not available, custom programs can be designed and combined with standard programs, as necessary, to accomplish the required control sequence.

DDC controllers also come with standard software for energy management programs. Among these programs are:

- Zero-energy band
- Load reset
- Optimum start
- Optimum stop
- Enthalpy control
- Duty cycle
- Night cycle
- Night purge
- Power demand limiting

These energy management programs can interact with the control loop programs, frequently using the same sensors and actuators. For instance, the control programs for a single zone HVAC unit can interact with the night cycle program. The unit can be cycled at night, at a reduced temperature, using the existing space sensor. During this cycling period, the existing outdoor air damper actuator can be maintained in the closed position. During summer operation, as determined by the existing outdoor air sensor, the fan can be cycled with cooling to maintain a maximum space temperature. Once again, the outdoor air damper actuator is maintained in the closed position to conserve energy. Many of the energy management programs listed in the previous paragraph would also apply to this HVAC unit using existing sensors and actuators. A layout of this system is shown in Figure 10.1.

Constructing a DDC System

Anyone who has designed a pneumatic or electric control system can construct a DDC system by following a few logical steps:

1. Make a schematic layout of the HVAC system to be controlled. Show all dampers, heating coils, cooling coils, fans, humidifiers, and any other pieces of mechanical equipment that are to be controlled. See Figure 10.2A.
2. Add actuators for the equipment to be controlled: damper actuators, control valves, relays, magnetic starters, contactors, and so forth. Next, add sensors, where required, to accomplish the anticipated control sequences. If additional sensors are required for energy management programs, add them at this time. See Figure 10.2B.
3. Enter the controller below the HVAC unit, and extend lines from all sensors and actuators to the controller. Label all input and output points at the controller according to their functions: digital input, analog input, digital output, analog output. See Figure 10.2C.
4. Compile a list of points connected to each controller, as shown in Figure 10.3. Note that DDC controllers will accept a limited number of input-output points. For this example (total 16 points), a 16 point controller can be used.

Sequence of Operation

This section describes the basic sequence of operation of a central air handling unit and a VAV box with DDC.

Central Air Handling Unit

The sequence of operation for the central air handling unit, as programmed into the system controller is as follows:

When the unit fan is energized, as sensed by the discharge duct static pressure sensor, the outdoor, return, and exhaust damper control system will be activated. A duct-mounted, discharge air temperature sensor will, through the system controller, maintain 55°F air supply to the remote VAV terminal units.

When the outdoor air temperature is below 65°F, as sensed by the temperature outdoor air sensor, the cooling coil valve will be in the coil bypass position and the discharge air temperature sensor will maintain temperature by modulating the heating coil valve in sequence with the outdoor, return, and exhaust air dampers.

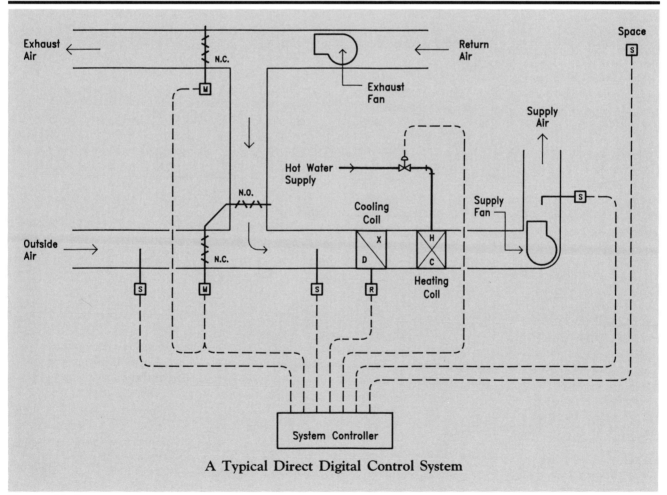

A Typical Direct Digital Control System

Figure 10.1

When the outdoor air temperature is above 65°F, the outdoor and exhaust air dampers will go to a preset minimum open position, and the return air damper will go to its corresponding open position. The discharge air sensor will now maintain air temperature by modulating the chilled water valve.

A low-temperature safety control, with its capillary located in the discharge side of the heating coil, will shut down the supply fan and return air fan and open the heating coil valve, if the heating coil discharge air temperature drops below its setting (35°F).

A static pressure sensor, located downstream from the supply fan, will maintain static pressure in the supply duct by modulating the supply air fan inlet vanes.

A ceiling-mounted static pressure sensor mounted in a representative location in the building will maintain a positive static pressure inside the building with reference to outdoors, by modulating the return air fan inlet vanes.

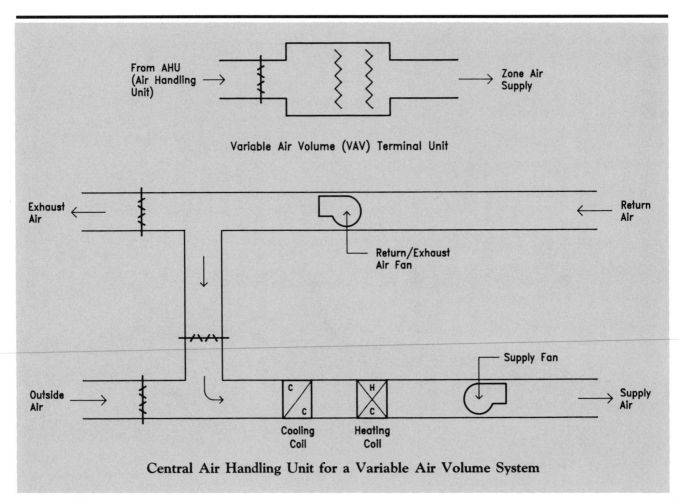

Central Air Handling Unit for a Variable Air Volume System

Figure 10.2a

Duct-mounted smoke detectors located in the supply and return air ducts will, upon sensing smoke, shut down the supply and return air fans, close the outdoor and exhaust dampers, and open the return air damper. (More complex smoke containment or evacuation sequences can be programmed if desired.)

During unoccupied periods in winter, the air handling unit fans will be de-energized, the outdoor and exhaust air dampers will close, and the return air damper will be in the full open position. If space temperature drops to the night setting (60°F) of a selected VAV terminal unit sensor (located in the coldest exposure), the unit fans will be cycled to maintain the reduced temperature. The outdoor and exhaust air dampers will remain closed; and the unit will operate on 100% return air. The heating coil valve will be open.

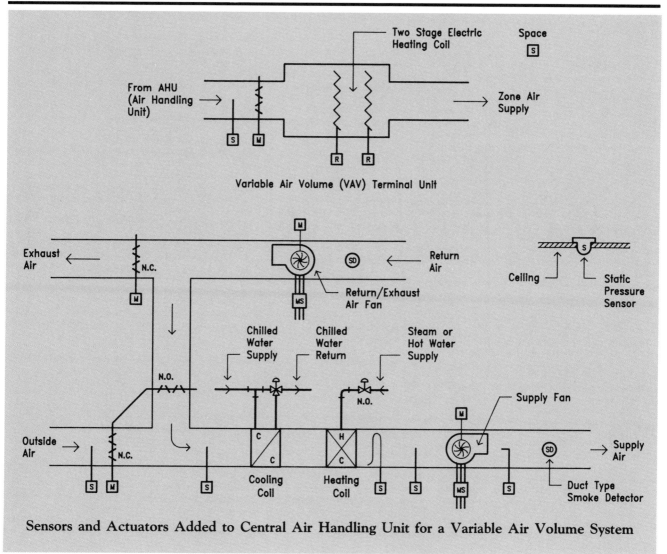

Sensors and Actuators Added to Central Air Handling Unit for a Variable Air Volume System

Figure 10.2b

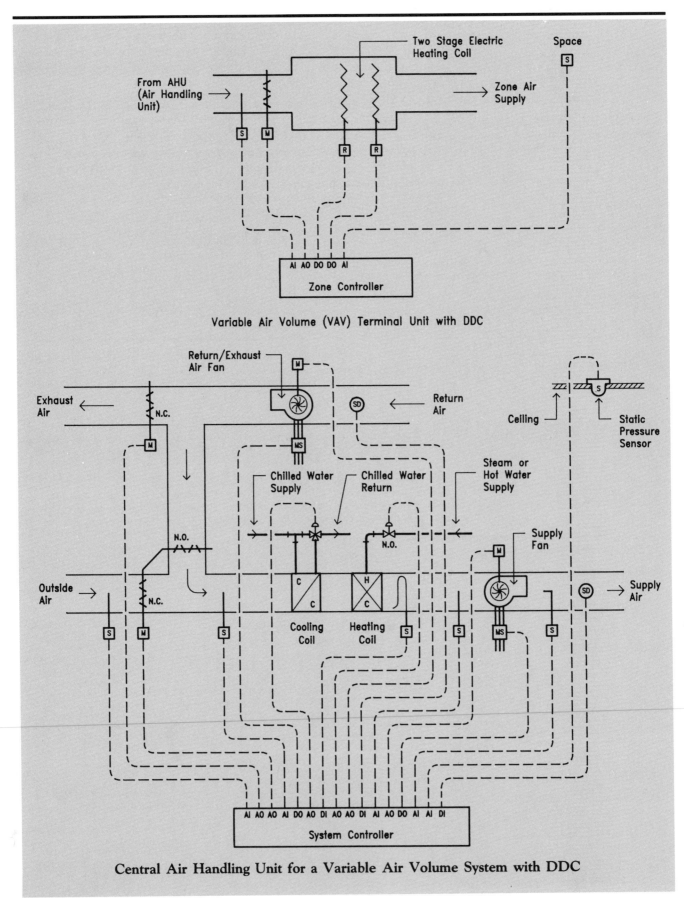

Variable Air Volume (VAV) Terminal Unit with DDC

Central Air Handling Unit for a Variable Air Volume System with DDC

Figure 10.2c

During unoccupied periods in summer, the air handling unit fans will be de-energized, the outdoor and exhaust air dampers will close, and the return air damper will be in the full open position. If space temperature rises to the night setting (80°F) of a selected VAV terminal unit sensor (located in the warmest exposure), the unit fans will be cycled to maintain the maximum temperature. The outdoor and exhaust dampers will remain closed, and the unit will operate on 100% return air. This program assumes the availability of cooling during unoccupied periods.

Whenever the unit supply fan is de-energized, as sensed by the discharge duct static pressure sensor, the return air fan will be de-energized; the outdoor and exhaust air dampers will close; and the return air damper will open.

Energy management programs resident in the DDC controller which could be applied to this air handling unit are:

- Optimum start/stop
- Load reset
- Night purge

Variable Air Volume Terminal Units

The sequence of operation for the variable air volume terminal units, as programmed into the zone controllers, is as follows:

On a call for less cooling by the space sensor, the variable air volume unit damper will modulate to the minimum position. The damper will remain at the minimum position through an adjustable deadband. If space temperature, as sensed by the space sensor, continues to drop, the stages of electric heat will be energized in sequence. When the heat is energized, the controller adjusts the unit air supply to protect the electric heating elements. On a call for cooling, the sequence is reversed.

The optional central host computer provides the capability to monitor, alarm, and adjust the set point of all sensor points as follows:

1. Outdoor air temperature
2. Mixed air temperature
3. Discharge air temperature
4. Supply duct static pressure
5. Space static pressure
6. Supply fan status (on-off-alarm)
7. Low temperature safety control (normal-off)
8. Smoke detectors (normal-alarm)
9. Space temperatures (total of 20)

Central Air Handling Unit	
AI	5
AO	6
DI	3
DO	2
Total Points	16

Point Summary

Figure 10.3

The CRT will display, either on request by the operator or automatically at preprogrammed intervals, the status of the above points. If a point deviates from its preprogrammed alarm limits, the CRT will display the point, its status or value, time and date, and action message directing the operator as to what action is to be taken. The printer will generate hard copy printouts of the information displayed on the CRT.

The maintenance program will schedule preventive maintenance and issue work orders based on calendar or equipment run time, or on an event occurrence. The CRT and printer will also display delinquent maintenance, inventory, and financial statistics pertaining to the HVAC system.

Budget Estimating

Following the four steps identified under the section *Constructing a DDC System* will simplify any system layout and make the budget estimating process more accurate.

For budget estimating purposes, we will use the central air handling unit with 20 VAV terminal units, as shown in Figure 10.2C. We will include an optional central host computer with color graphics and basic maintenance manager software. Average wiring runs for points in mechanical rooms are estimated at 150 feet, and to space temperature sensors, 150 feet. Cost information is found in *Means Mechanical Cost Data* (HVAC/controls), Division 157-400 (420 and 422). See Figure 10.4. The 1991 edition was used for this exercise. See Figure 10.5 for the sample budget estimating procedures.

An accurate budget estimate requires discipline and a structured approach to system layout. Following the procedures discussed in this chapter will simplify the system layout and budget estimating process.

			DAILY	MAN-		BARE COSTS				TOTAL	
	157 400	**Accessories**	CREW	OUTPUT	HOURS	UNIT	MAT.	LABOR	EQUIP.	TOTAL	INCL O&P
401	1870	Inlet or outlet transition, vertical										**401**
	1880	Single pack to 5000 CFM, add					2%					
	1890	Single pack to 24,000 CFM, add					3%					
	1900	Double pack to 24,000 CFM, add					2%					
	2000	Electronic air cleaner, self-contained										
	2050	185 CFM	1 Shee	2.40	3.330	Ea.	211	83		294	360	
	2100	200 CFM		2.30	3.480		410	87		497	585	
	2150	500 CFM		2.30	3.480		585	87		672	775	
	2200	1000 CFM		2.20	3.640		810	91		901	1,025	
	2250	1200 CFM		2.10	3.810		1,190	95		1,285	1,450	
	2300	2500 CFM	↓	2	4	↓	1,360	100		1,460	1,650	
	2400	In-line installation, duct or plenum										
	2420	Power pack, one required for each										
	2430	Installation, one or two filter cells	1 Shee	2	4	Ea.	221	100		321	395	
	2440	Filter cell, 2" thick, 1" perimeter offset		4	2		215	50		265	310	
	2450	20" x 20"		4	2		170	50		220	265	
	2460	20" x 16"		4	2		152	50		202	245	
	2470	25" x 20"		3.50	2.290		193	57		250	300	
	2480	25" x 16"	↓	3	2.670	↓	170	67		237	290	
	2950	Mechanical media filtration units										
	3000	High efficiency type, with frame, non-supported				MCFM	35			35	39	
	3100	Supported type					47			47	52	
	4000	Medium efficiency, extended surface					5			5	5.50	
	4500	Permanent washable					33			33	36	
	5000	Renewable disposable roll				↓	85			85	94	
	5500	Throwaway glass or paper media type				Ea.	2.65			2.65	2.92	
410	0010	**ANTI-FREEZE** Inhibited										**410**
	0900	Ethylene glycol concentrated										
	1000	55 gallon drums, small quantities				Gal.	6.71			6.71	7.40	
	1200	Large quantities					6.15			6.15	6.75	
	2000	Propylene glycol, for solar heat, small quantities					7.39			7.39	8.15	
	2100	Large quantities				↓	6.74			6.74	7.40	
420	0010	**CONTROL COMPONENTS**										**420**
	0700	Controller, receiver										
	0730	Pneumatic, panel mount, single input	1 Plum	8	1	Ea.	183	25		208	240	
	0740	With conversion mounting bracket		8	1		188	25		213	245	
	0750	Dual input, with control point adjustment	↓	7	1.140		206	29		235	270	
	0850	Electric, single snap switch	1 Elec	4	2		239	50		289	335	
	0860	Dual snap switches		3	2.670		317	67		384	450	
	0870	Humidity controller		8	1		139	25		164	190	
	0880	Load limiting controller		8	1		260	25		285	325	
	0890	Temperature controller	↓	8	1	↓	125	25		150	175	
	1000	Enthalpy control, boiler water temperature control										
	1010	governed by outdoor temperature, with timer	1 Elec	3	2.670	Ea.	125	67		192	235	
	2000	Gauges, pressure or vacuum										
	2100	2" diameter dial	1 Stpi	32	.250	Ea.	8	6.40		14.40	18.30	
	2200	2-½" diameter dial		32	.250		9.40	6.40		15.80	19.85	
	2300	3-½" diameter dial		32	.250		12	6.40		18.40	23	
	2400	4-½" diameter dial	↓	32	.250	↓	16.50	6.40		22.90	28	
	2700	Flanged iron case, black ring										
	2800	3-½" diameter dial	1 Stpi	32	.250	Ea.	42.75	6.40		49.15	57	
	2900	4-½" diameter dial		32	.250		51.75	6.40		58.15	66	
	3000	6" diameter dial	↓	32	.250	↓	70.50	6.40		76.90	87	
	3300	For compound pressure-vacuum, add					18%					
	3350	Humidistat										
	3360	Pneumatic operation										
	3361	Room humidistat, direct acting	1 Stpi	12	.667	Ea.	108	17		125	145	
	3362	Room humidistat, reverse acting	"	12	.667	"	108	17		125	145	

245

Figure 10.4

		157 400	Accessories	CREW	DAILY OUTPUT	MAN-HOURS	UNIT	BARE COSTS				TOTAL INCL O&P	
								MAT.	LABOR	EQUIP.	TOTAL		
420	3363		Room humidity transmitter	1 Stpi	17	.471	Ea.	156	12		168	190	420
	3364		Duct mounted controller		12	.667		152	17		169	195	
	3365		Duct mounted transmitter		12	.667		152	17		169	195	
	3366		Humidity indicator, 3-½"		28	.286		57	7.30		64.30	74	
	3390		Electric operated	1 Shee	8	1		35.75	25		60.75	77	
	3400		Relays										
	3430		Pneumatic/electric	1 Plum	16	.500	Ea.	255	12.75		267.75	300	
	3440		Pneumatic proportioning		8	1		120	25		145	170	
	3450		Pneumatic switching		12	.667		61	16.95		77.95	92	
	3460		Selector, 3 point		6	1.330		42.80	34		76.80	98	
	3470		Time delay		8	1		120	25		145	170	
	3500		Sensor, air operated										
	3520		Humidity	1 Plum	16	.500	Ea.	164	12.75		176.75	200	
	3540		Pressure		16	.500		103	12.75		115.75	130	
	3560		Temperature		12	.667		118	16.95		134.95	155	
	3600		Electric operated										
	3620		Humidity	1 Elec	8	1	Ea.	40.40	25		65.40	82	
	3650		Pressure		8	1		485	25		510	570	
	3680		Temperature		10	.800		69.20	20		89.20	105	
	4000		Thermometers										
	4100		Dial type, 3-½" diameter, vapor type, union connection	1 Stpi	32	.250	Ea.	95	6.40		101.40	115	
	4120		Liquid type, union connection		32	.250		133	6.40		139.40	155	
	4130		Remote reading, 15' capillary		32	.250		110	6.40		116.40	130	
	4500		Stem type, 6-½" case, 2" stem, ½" NPT		32	.250		24	6.40		30.40	36	
	4520		4" stem, ½" NPT		32	.250		33	6.40		39.40	46	
	4600		9" case, 3-½" stem, ¾" NPT		28	.286		40	7.30		47.30	55	
	4620		6" stem, ¾" NPT		28	.286		43	7.30		50.30	58	
	4640		8" stem, ¾" NPT		28	.286		49	7.30		56.30	65	
	4660		12" stem, 1" NPT		26	.308		55	7.85		62.85	72	
	5000		Thermostats										
	5030		1 set back, manual	1 Shee	8	1	Ea.	16.30	25		41.30	56	
	5040		1 set back, electric, timed		8	1		59.45	25		84.45	105	
	5050		2 set back, electric, timed		8	1		59.45	25		84.45	105	
	5100		Locking cover					13.90			13.90	15.30	
	5200		24 hour, automatic, clock	1 Shee	8	1		86.90	25		111.90	135	
	5220		Electric, 2 wire	1 Elec	13	.615		12.35	15.50		27.85	36	
	5230		3 wire	"	10	.800		14.95	20		34.95	46	
	5240		Pneumatic										
	5250		Single temp., single pressure	1 Stpi	8	1	Ea.	115	26		141	165	
	5251		Dual pressure		8	1		159	26		185	215	
	5252		Dual temp., dual pressure		8	1		155	26		181	210	
	5253		Reverse acting w/averaging element		8	1		112	26		138	160	
	5254		Heating-cooling w/deadband		8	1		155	26		181	210	
	5255		Integral w/piston top valve actuator		8	1		144	26		170	195	
	5256		Dual temp., dual pressure		8	1		164	26		190	220	
	5257		Low limit, 8' averaging element		8	1		91	26		117	140	
	5258		Room single temp. proportional		8	1		58.30	26		84.30	100	
	5300		Transmitter, pneumatic										
	5320		Temperature averaging element	Q-1	8	2	Ea.	79.80	46		125.80	155	
	5350		Pressure differential	1 Plum	7	1.140		438	29		467	525	
	5370		Humidity, duct		8	1		152	25		177	205	
	5380		Room		12	.667		156	16.95		172.95	195	
	5390		Temperature, with averaging element		6	1.330		104	34		138	165	
	5420		Electric operated, humidity	1 Elec	8	1		156	25		181	210	
	5430		DPST	"	8	1		35.50	25		60.50	76	
	6000		Valves, motorized zone										
	6100		Sweat connections, ½" C x C	1 Stpi	20	.400	Ea.	34.10	10.20		44.30	53	
	6110		¾" C x C	"	20	.400	"	34.10	10.20		44.30	53	

246

Figure 10.4 *(continued)*

			DAILY	MAN-		BARE COSTS				TOTAL	
	157 400 |Accessories	CREW	OUTPUT	HOURS	UNIT	MAT.	LABOR	EQUIP.	TOTAL	INCL O&P	
420	6120	1″ C x C	1 Stpi	19	.421	Ea.	38.15	10.75		48.90	58	**420**
	6140	½″ C x C, with end switch, 2 wire		20	.400		41.20	10.20		51.40	61	
	6150	¾″ C x C, with end switch, 2 wire		20	.400		43.10	10.20		53.30	63	
	6160	1″ C x C, with end switch, 2 wire		19	.421		47.90	10.75		58.65	69	
	6180	1-¼″ C x C, w/end switch, 2 wire		15	.533		52.90	13.60		66.50	78	
	7090	Valves, motor controlled, including actuator										
	7100	Electric motor actuated										
	7200	Brass, two way, screwed										
	7210	½″ pipe size	L-6	36	.333	Ea.	163	8.45		171.45	190	
	7220	¾″ pipe size		30	.400		182	10.15		192.15	215	
	7230	1″ pipe size		28	.429		201	10.85		211.85	235	
	7240	1-½″ pipe size		19	.632		279	16		295	330	
	7250	2″ pipe size		16	.750		425	19		444	495	
	7350	Brass, three way, screwed										
	7360	½″ pipe size	L-6	33	.364	Ea.	204	9.20		213.20	240	
	7370	¾″ pipe size		27	.444		230	11.25		241.25	270	
	7380	1″ pipe size		25.50	.471		300	11.95		311.95	350	
	7390	1-½″ pipe size		17	.706		306	17.90		323.90	365	
	7400	2″ pipe size		14	.857		460	22		482	540	
	7550	Iron body, two way, flanged										
	7560	2-½″ pipe size	L-6	4	3	Ea.	535	76		611	700	
	7570	3″ pipe size		3	4		625	100		725	840	
	7580	4″ pipe size		2	6		890	150		1,040	1,200	
	7590	6″ pipe size		1.50	8		2,490	205		2,695	3,050	
	7850	Iron body, three way, flanged										
	7860	2-½″ pipe size	L-6	3	4	Ea.	800	100		900	1,025	
	7870	3″ pipe size		2.50	4.800		1,030	120		1,150	1,325	
	7880	4″ pipe size		2	6		1,180	150		1,330	1,525	
	7890	6″ pipe size		1.50	8		3,060	205		3,265	3,675	
	8000	Pneumatic, air operated										
	8050	Brass, two way, screwed										
	8060	½″ pipe size, class 250	1 Plum	24	.333	Ea.	112	8.50		120.50	135	
	8070	¾″ pipe size, class 250		20	.400		127	10.20		137.20	155	
	8080	1″ pipe size, class 250		19	.421		153	10.70		163.70	185	
	8090	1-¼″ pipe size, class 125		15	.533		215	13.55		228.55	255	
	8100	1-½″ pipe size, class 125		13	.615		224	15.65		239.65	270	
	8110	2″ pipe size, class 125		11	.727		330	18.50		348.50	390	
	8180	Brass, three way, screwed										
	8190	½″ pipe size, class 250	1 Plum	22	.364	Ea.	135	9.25		144.25	160	
	8200	¾″ pipe size, class 250		18	.444		140	11.30		151.30	170	
	8210	1″ pipe size, class 250		17	.471		176	12		188	210	
	8220	1-½″ pipe size, class 125		11	.727		256	18.50		274.50	310	
	8230	2″ pipe size, class 125		9	.889		345	23		368	415	
	8250	Iron body, two way, screwed										
	8260	½″ pipe size, class 300	1 Plum	24	.333	Ea.	498	8.50		506.50	560	
	8270	¾″ pipe size, class 300		20	.400		510	10.20		520.20	575	
	8280	1″ pipe size, class 300		19	.421		695	10.70		705.70	780	
	8290	1-¼″ pipe size, class 300		15	.533		990	13.55		1,003	1,100	
	8300	1-½″ pipe size, class 300		13	.615		1,010	15.65		1,025	1,125	
	8310	2″ pipe size, class 300		11	.727		1,010	18.50		1,028	1,150	
	8320	1-¼″ pipe size, class 125		15	.533		285	13.55		298.55	335	
	8330	1-½″ pipe size, class 125		13	.615		317	15.65		332.65	370	
	8340	2″ pipe size, class 125		11	.727		395	18.50		413.50	460	
	8450	Iron body, two way, flanged										
	8460	2-½″ pipe size, 250 lb. flanges	Q-1	5	3.200	Ea.	2,060	73		2,133	2,375	
	8470	3″ pipe size, 250 lb. flanges		4.50	3.560		2,200	81		2,281	2,550	
	8480	4″ pipe size, 250 lb. flanges		3	5.330		2,530	120		2,650	2,975	
	8510	2-½″ pipe size, class 125		5	3.200		605	73		678	775	

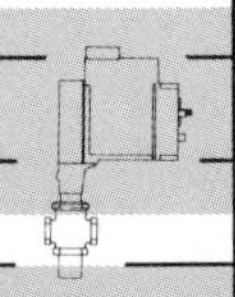

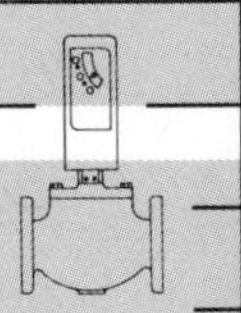

247

Figure 10.4 *(continued)*

157 400	Accessories	CREW	DAILY OUTPUT	MAN-HOURS	UNIT	BARE COSTS				TOTAL INCL O&P	
						MAT.	LABOR	EQUIP.	TOTAL		
420	8520	3" pipe size, class 125	Q-1	4.50	3.560	Ea.	845	81		926	1,050	420
	8530	4" pipe size, class 125	"	3	5.330		1,050	120		1,170	1,325	
	8540	5" pipe size, class 125	Q-2	3.40	7.060		1,720	170		1,890	2,150	
	8550	6" pipe size, class 125	"	3	8		1,850	190		2,040	2,325	
	8560	Iron body, three way, flanged										
	8570	2-½" pipe size, class 125	Q-1	4.50	3.560	Ea.	760	81		841	960	
	8580	3" pipe size, class 125		4	4		915	92		1,007	1,150	
	8590	4" pipe size, class 125		2.50	6.400		1,490	145		1,635	1,850	
	8600	6" pipe size, class 125	Q-2	3	8		2,230	190		2,420	2,725	
	9005	Pneumatic system misc. components										
	9010	Adding/subtracting repeater	1 Stpi	20	.400	Ea.	42.25	10.20		52.45	62	
	9020	Adjustable ratio network		16	.500		124	12.75		136.75	155	
	9030	Comparator		20	.400		126	10.20		136.20	155	
	9040	Cumulator										
	9041	Air switching	1 Stpi	20	.400	Ea.	61	10.20		71.20	82	
	9042	Averaging		20	.400		65	10.20		75.20	87	
	9043	2:1 ratio		20	.400		56	10.20		66.20	77	
	9044	Sequencing		20	.400		38	10.20		48.20	57	
	9045	Two-position		16	.500		65	12.75		77.75	91	
	9046	Two-position pilot		16	.500		61	12.75		73.75	86	
	9050	Damper actuator										
	9051	For smoke control	1 Stpi	8	1	Ea.	117	26		143	165	
	9052	For ventilation	"	8	1	"	181	26		207	235	
	9053	Series duplex pedestal mounted										
	9054	For inlet vanes on fans, compressors	1 Stpi	6	1.330	Ea.	610	34		644	720	
	9060	Enthalpy logic center		16	.500		128	12.75		140.75	160	
	9070	High/low pressure selector		22	.364		27	9.25		36.25	44	
	9080	Signal limiter		20	.400		38.25	10.20		48.45	57	
	9081	Transmitter		28	.286		15	7.30		22.30	27	
	9090	Optimal start										
	9091	Warmup or cooldown programmer	1 Stpi	10	.800	Ea.	448	20		468	525	
	9092	Mass temperature transmitter	"	18	.444	"	114	11.35		125.35	140	
	9100	Step controller with positioner, time delay restrictor										
	9110	recycler air valve, switches and cam settings										
	9112	With cabinet										
	9113	6 points	1 Stpi	4	2	Ea.	330	51		381	440	
	9114	6 points, 6 manual sequences		2	4		348	100		448	535	
	9115	8 points		3	2.670		418	68		486	560	
	9200	Pressure controller and switches										
	9210	High static pressure limit	1 Stpi	8	1	Ea.	97	26		123	145	
	9220	Pressure transmitter		8	1		105	26		131	155	
	9230	Differential pressure transmitter		8	1		438	26		464	520	
	9240	Static pressure transmitter		8	1		268	26		294	335	
	9250	Proportional-only control, single input		6	1.330		183	34		217	250	
	9260	Proportional plus integral, single input		6	1.330		255	34		289	330	
	9270	Proportional-only, dual input		6	1.330		206	34		240	275	
	9280	Proportional plus integral, dual input		6	1.330		325	34		359	410	
	9281	Time delay for above proportional units		16	.500		72	12.75		84.75	98	
	9290	Proportional/2 position controller units		6	1.330		214	34		248	285	
	9300	Differential press. control dir./rev.		6	1.330		325	34		359	410	
	9310	Air pressure reducing valve										
	9311	⅛" size	1 Stpi	18	.444	Ea.	14.50	11.35		25.85	33	
	9312	⅜" size		17	.471		27	12		39	48	
	9313	½" size		16	.500		34	12.75		46.75	56	
	9314	¾" size		14	.571		60	14.55		74.55	88	
	9315	Precision valve, ¼" size		17	.471		82.50	12		94.50	110	
	9320	Air flow controller		12	.667		83	17		100	115	
	9330	Booster relay volume amplifier		18	.444		45.75	11.35		57.10	67	

248

Figure 10.4 (continued)

			DAILY	MAN-		BARE COSTS				TOTAL	
157 400	**Accessories**	CREW	OUTPUT	HOURS	UNIT	MAT.	LABOR	EQUIP.	TOTAL	INCL O&P	
420	9331	Reverse acting	1 Stpi	18	.444	Ea.	86	11.35		97.35	110	420
	9340	Series restrictor, straight		60	.133		2.45	3.40		5.85	7.75	
	9341	T-fitting		40	.200		3.60	5.10		8.70	11.55	
	9342	In-line adjustable		48	.167		15.50	4.25		19.75	23	
	9350	Diode tee		40	.200		8.25	5.10		13.35	16.70	
	9351	Restrictor tee		40	.200		8.25	5.10		13.35	16.70	
	9360	Pneumatic gradual switch		8	1		100	26		126	150	
	9361	Selector switch		8	1		48.25	26		74.25	91	
	9370	Electro-pneumatic motor driven servo		10	.800		408	20		428	480	
	9380	Regulated power supply		10	.800		350	20		370	415	
	9390	Fan control switch and mounting base		16	.500		42.25	12.75		55	66	
	9400	Circulating pump sequencer	↓	14	.571	↓	455	14.55		469.55	520	
	9410	Pneumatic tubing, fittings and accessories										
	9414	Tubing, urethane										
	9415	⅛″ OD x 1⁄16″ ID	1 Stpi	120	.067	L.F.	.34	1.70		2.04	2.91	
	9416	¼″ OD x ⅛″ ID		115	.070		.64	1.77		2.41	3.35	
	9417	5⁄32″ OD x 3⁄32″ ID	↓	110	.073	↓	.36	1.85		2.21	3.16	
	9420	Coupling, straight										
	9422	Barb x barb										
	9423	¼″ x 5⁄32″	1 Stpi	160	.050	Ea.	.21	1.28		1.49	2.13	
	9424	¼″ x ¼″		158	.051		.29	1.29		1.58	2.25	
	9425	⅜″ x ⅜″		154	.052		.26	1.32		1.58	2.26	
	9426	⅜″ x ¼″		150	.053		.21	1.36		1.57	2.26	
	9427	½″ x ¼″		148	.054		.54	1.38		1.92	2.65	
	9428	½″ x ⅜″		144	.056		.45	1.42		1.87	2.61	
	9429	½″ x ½″	↓	140	.057	↓	.27	1.46		1.73	2.47	
	9440	Tube x tube										
	9441	¼″ x ¼″	1 Stpi	100	.080	Ea.	.92	2.04		2.96	4.06	
	9442	⅜″ x ¼″		96	.083		1.03	2.13		3.16	4.30	
	9443	⅜″ x ⅜″		92	.087		1.03	2.22		3.25	4.44	
	9444	½″ x ⅜″		88	.091		1.65	2.32		3.97	5.25	
	9445	½″ x ½″	↓	84	.095		2.11	2.43		4.54	5.95	
	9450	Elbow coupling										
	9452	Barb x barb										
	9454	¼″ x ¼″	1 Stpi	158	.051	Ea.	.52	1.29		1.81	2.50	
	9455	⅜″ x ⅜″		154	.052		.50	1.32		1.82	2.53	
	9456	½″ x ½″	↓	140	.057		.83	1.46		2.29	3.09	
	9460	Tube x tube										
	9462	½″ x ½″	1 Stpi	84	.095	Ea.	2.01	2.43		4.44	5.85	
	9470	Tee coupling										
	9472	Barb x barb x barb										
	9474	¼″ x ¼″ x 5⁄32″	1 Stpi	108	.074	Ea.	.45	1.89		2.34	3.31	
	9475	¼″ x ¼″ x ¼″		104	.077		.54	1.96		2.50	3.52	
	9476	⅜″ x ⅜″ x 5⁄32″		102	.078		2.32	2		4.32	5.55	
	9477	⅜″ x ⅜″ x ¼″		99	.081		.88	2.06		2.94	4.04	
	9478	⅜″ x ⅜″ x ⅜″		98	.082		1.03	2.08		3.11	4.24	
	9479	½″ x ½″ x ¼″		96	.083		.83	2.13		2.96	4.08	
	9480	½″ x ½″ x ⅜″		95	.084		.83	2.15		2.98	4.12	
	9481	½″ x ½″ x ½″	↓	92	.087	↓	.83	2.22		3.05	4.22	
	9484	Tube x tube x tube										
	9485	¼″ x ¼″ x ¼″	1 Stpi	66	.121	Ea.	1.70	3.09		4.79	6.50	
	9486	⅜″ x ¼″ x ¼″		64	.125		3.04	3.19		6.23	8.10	
	9487	⅜″ x ⅜″ x ¼″		61	.131		2.63	3.34		5.97	7.90	
	9488	⅜″ x ⅜″ x ⅜″		60	.133		2.37	3.40		5.77	7.70	
	9489	½″ x ½″ x ⅜″		58	.138		4.07	3.52		7.59	9.70	
	9490	½″ x ½″ x ½″	↓	56	.143	↓	4.07	3.64		7.71	9.90	
	9492	Needle valve										
	9494	Tube x tube										

249

Figure 10.4 (continued)

| | | | DAILY | MAN- | | BARE COSTS | | | | TOTAL | |
157 400	**Accessories**	CREW	OUTPUT	HOURS	UNIT	MAT.	LABOR	EQUIP.	TOTAL	INCL O&P	
420 9495	¼″ x ¼″	1 Stpi	86	.093	Ea.	2.58	2.37		4.95	6.40	**420**
9496	⅜″ x ⅜″	S-TPI	82		″	3.20			3.20	3.52	
9600	Electronic system misc. components										
9610	Electric motor damper actuator	1 Elec	8	1	Ea.	182	25		207	235	
9620	Damper position indicator		10	.800		63.50	20		83.50	100	
9630	Pneumatic-electronic transducer		12	.667		204	16.75		220.75	250	
9640	Electro-pneumatic transducer		12	.667		71.60	16.75		88.35	105	
9650	Remote reset control with motor drive		6	1.330		166	34		200	230	
9660	Manual remote set point control		16	.500		32.70	12.60		45.30	55	
9670	Alarm unit with two adjustable points	↓	12	.667	↓	140	16.75		156.75	180	
9700	Solid state heater step controller										
9701	4 stage	1 Elec	10	.800	Ea.	256	20		276	310	
9702	6 stage	″	8	1	″	320	25		345	390	
9710	Staging network										
9711	Master one stage	1 Elec	10	.800	Ea.	92.70	20		112.70	130	
9712	Two stage		9	.889	↓	155	22		177	205	
9720	Sequencing network		8	1		164	25		189	220	
9730	Regulated DC power supply	↓	16	.500	↓	135	12.60		147.60	165	
9740	Accessory unit regulator										
9741	Three function	1 Elec	9	.889	Ea.	83.50	22		105.50	125	
9742	Five function		8	1		101	25		126	150	
9750	Accessory power supply		18	.444		26.80	11.20		38	46	
9760	Step down transformer	↓	16	.500	↓	107	12.60		119.60	135	
422 0010	**CONTROL COMPONENTS / DDC SYSTEMS** (Sub's quote incl. M & L)										**422**
0100	Analog inputs										
0110	Sensors (avg. 150′ run in conduit)										
0120	Duct temperature				Ea.					340	
0130	Space temperature									610	
0140	Duct humidity, +/- 3%									640	
0150	Space humidity, +/- 2%									980	
0160	Duct static pressure									520	
0170	C.F.M./Transducer									700	
0172	Water temp. (see 156-210 for well tap add)									600	
0174	Water flow (see 156-210 for circuit sensor add)									2,200	
0176	Water pressure differential (see 156-210 for tap add)									900	
0177	Steam flow (see 156-210 for circuit sensor add)									2,200	
0178	Steam pressure (see 156-210 for tap add)									940	
0180	K.W./Transducer									1,250	
0182	K.W.H. totalization (not incl. elec. meter pulse xmte.)									575	
0190	Space static pressure				↓					1,000	
1000	Analog outputs (avg. 50′ run in conduit)										
1010	P/I Transducer				Ea.					580	
1020	Analog output, matl. in MUX									280	
1030	Pneumatic (not incl. control device)									590	
1040	Electric (not incl control device)				↓					350	
2000	Status (Alarms)										
2100	Digital inputs										
2110	Freeze				Ea.					400	
2120	Fire									360	
2130	Differential pressure, (air)									550	
2140	Differential pressure, (water)									780	
2150	Current sensor									400	
2170	Duct smoke detector									650	
2200	Digital output										
2210	Start/stop				Ea.					320	
2220	On/off (maintained contact)				″					550	
3000	Controller m.U.X. panel, incl. function boards										

250

Figure 10.4 (continued)

			CREW	DAILY OUTPUT	MAN-HOURS	UNIT	BARE COSTS				TOTAL INCL O&P	
	157 400	**Accessories**					MAT.	LABOR	EQUIP.	TOTAL		
422	3100	48 point				Ea.					4,850	422
	3110	128 point				"					6,650	
	3200	D.D.C. controller (avg. 50' run in conduit)										
	3210	Mechanical room										
	3214	16 point controller (incl. 120v/1ph power supply)				Ea.					3,000	
	3229	32 point controller (incl. 120v/1ph power supply)				"					5,000	
	3230	Includes software programming and checkout										
	3260	Space										
	3266	V.A.V. terminal box (incl. space terp. sensor)				Ea.					775	
	3280	Host computer (avg. 50' run in conduit)										
	3281	Package complete with 386 PC, keyboard										
	3282	printer, color CRT, modem, basic software				Ea.					15,000	
	4000	Front end costs										
	4100	Computer (P.C.)/software program				Ea.					6,000	
	4200	Color graphics software									3,600	
	4300	Color graphics slides									450	
	4350	Additional dot matrix printer				↓					900	
	4400	Communications trunk cable				Ft					3.50	
	4500	Engineering labor, (not incl. dftg.)				Point					76	
	4600	Calibration labor									76	
	4700	Start-up, checkout labor				↓					115	
	4800	Drafting labor, as req'd										
	5000	Communications bus (data transmission cable)										
	5010	#18 twisted shielded pair in conduit				C.L.F.					850	
	8000	Applications software										
	8050	Basic maintenance manager software (not incl. data base entry)				Ea.					1,800	
	8100	Time program				Point					6.30	
	8120	Duty cycle									12.55	
	8140	Optimum start/stop									38	
	8160	Demand limiting									18.80	
	8180	Enthalpy program				↓					38	
	8200	Boiler optimization				Ea.					1,125	
	8220	Chiller optimization				"					1,500	
	8240	Custom applications										
	8260	Cost varies with complexity										
425	0010	**CONTROL SYSTEMS, PNEUMATIC** (Sub's quote incl. mat. & labor)										425
	0100	Heating and Ventilating, split system										
	0200	Mixed air control, economizer cycle, panel readout, tubing										
	0220	Up to 10 tons				Ea.					3,650	
	0240	For 10 to 20 tons, add [C8.3 -701]									8%	
	0260	For over 20 tons, add									17%	
	0270	Enthalpy cycle, up to 10 tons									4,425	
	0280	For 10 to 20 tons, add									8%	
	0290	For over 20 tons, add				↓					17%	
	0300	Heating coil, hot water, 3 way valve,										
	0320	freezestat, limit control on discharge readout				Ea.					2,625	
	0500	Cooling coil, chilled water, room										
	0520	thermostat, 3 way valve				Ea.					1,100	
	0600	Cooling tower, fan cycle, damper control,										
	0620	condenser, water readout in/out at panel				Ea.					4,575	
	1000	Unit ventilator, day/night operation,										
	1100	freezestat, ASHRAE, cycle 2				Ea.					2,700	
	2000	Compensated hot water from boiler, valve control,										
	2100	readout and reset at panel, up to 60 GPM				Ea.					4,525	
	2120	For 120 GPM, add									7%	
	2140	For 240 GPM, add				↓					12%	
	3000	Boiler room combustion air, damper, controls									2,050	

251

Figure 10.4 (continued)

Means Forms
CONSOLIDATED ESTIMATE

PROJECT **BAS Budget Estimate** CLASSIFICATION **HVAC/Controls**

LOCATION **Kingston, Ma.**

ARCHITECT **R. Carlson**

TAKE OFF BY **RD** QUANTITIES BY **RD** PRICES BY **RC** EXTENSIONS BY **JC** DATE **1-4-91** CHECKED BY **MG**

Description	Quantities		Installed Cost	Total Cost
	Budget	Unit		
Central Air Handling Unit				
Control Components (Electric)				
Damper Actuator (O.A.-R.A.)	1	Ea.	235	235
Damper Actuator (E.A.)	1	Ea.	235	235
Cooling Coil Control Valve - 2"	1	Ea.	540	540
Heating Coil Control Valve 1½"	1	Ea.	330	330
Damper Actuator (Supply Fan Vortex)	1	Ea.	720	720
Damper Actuator (Return Fan Vortex)	1	Ea.	720	720
			Total	2780
Control Components/DDC Systems				
Outdoor Air Temperature Sensor (A.I.)	1	Ea.	340	340
Damper Actuator O.A.-R.A. (A.O.)	1	Ea.	350	350
Damper Actuator E.A. (A.O.)	1	Ea.	350	350
Mixed Air Temperature Sensor (A.I.)	1	Ea.	340	340
Cooling Coil Control Valve (A.O.)	1	Ea.	350	350
Heating Coil Control Valve (A.O.)	1	Ea.	350	350
Low Temperature Detector (D.I.)	1	Ea.	400	400
Discharge Air Temperature Sensor (A.I.)	1	Ea.	340	340
Damper Actuator Supply Fan Vortex (A.O.)	1	Ea.	350	350
Damper Actuator Return Fan Vortex (A.O.)	1	Ea.	350	350
Supply Duct Static Pressure Sensor (A.I.)	1	Ea.	520	520
Space Static Pressure Sensor (A.I.)	1	Ea.	1000	1000

Figure 10.5

174

Means Forms
CONSOLIDATED ESTIMATE

PROJECT BAS Budget Estimate

LOCATION Kingston, Ma

CLASSIFICATION HVAC/controls

ARCHITECT R. Carlson

TAKE OFF BY RD QUANTITIES BY RD PRICES BY RC EXTENSIONS BY JC

Description	Quantities Budget	Unit	Installed Cost	Total Cost
Control Components/DDC Systems (cont'd)				
Duct Smoke Detector - Return Air (D.I.)	1	Ea.	650	650
Duct Smoke Detector - Supply Air (D.I.)	1	Ea.	650	650
Supply Fan Start/Stop	1	Ea.	320	320
Return Fan Start/Stop	1	Ea.	320	320
DDC System Controller (16 Point)	1	Ea.	3000	3000
			Total	9980
V.A.V. Terminal Units (Total of 20)				
V.A.V. Terminal Box (includes space temperature sensor)	20 Ea.		775	15500
			Total	15500
Optional Central Host Computer				
386 PC, Keyboard, Printer, Color CRT, modem, Basic Software	1 Ea.		15000	15000
Colorgraphic Software	1 Ea.		3600	3600
Individual Colorgraphics	6 Ea.		450	2700
Maintenance Manager Software	1 Ea.		1800	1800
			Total	23100
Communications Bus				
Communications Bus (#18 Twisted Shielded Pair in Enclosure)	500 Ft.		850/100 Ft.	4250
			Total	4250
Grand Total				55610

Figure 10.5 (continued)

Part Four

System Selection Criteria

When a decision is made to install a Building Automation System (BAS), certain steps should be taken and criteria established for the selection process. If a BAS is to be an effective tool for a building owner and his operating staff, a detailed review of available systems, their features, and capabilities is a must, prior to making a buying decision.

The review should be conducted by a team, comprised of the building owner or his/her representative, a financial advisor, a consulting engineer familiar with BAS, and a selected member of the building operating staff. The team should be assigned the responsibility of establishing the goals and objectives to be achieved by installing a BAS. Team members must follow a logical, structured approach in evaluating systems, with an emphasis on economy, compatibility with building systems, user-friendliness, and vendor support quality.

In the following chapter we will offer some suggestions on what criteria to use in evaluating Building Automation Systems.

Chapter Eleven

System Selection Criteria

Why select a Building Automation System? A BAS is a powerful tool for building owners, managers, and operating personnel. It can provide accurate temperature control to ensure the optimum in comfort conditions and building air quality. Energy management programs are available to assist operating personnel in the most efficient utilization of energy-consuming systems. Continuous monitoring of building systems can assist the maintenance staff by providing early notification in the event of equipment failure. An operator can also predict equipment failure by observing ongoing trend reports. Preventive maintenance scheduling, energy audits, and many other building management reports are other important tools that a BAS provides for the day-to-day process of managing a building.

Building Automation Systems are designed to integrate building systems, such as HVAC, lighting, fire safety, and security/access control. The integration of all of these systems can reduce initial and ongoing costs, and simplify building operation. In addition, interactive software programs can improve or enhance life safety and property protection. For example, fire safety software can interact with HVAC programs for smoke control. Fire safety software can also interact with door access programs to provide a means of egress for building occupants in the event of a fire. Both are advantages that improve the life safety conditions of a building.

The aforementioned, and many other features of BAS, provide major benefits to the building owner. They enhance comfort, energy management, life safety, and asset protection, which are the ultimate goals of every building owner or manager.

Although many manufacturers compete in the business of BAS, 80% of the market is virtually controlled by four or five of the vendors. There is, however, an opportunity to shop for the "best deal." It should be noted that the "best deal" is not necessarily the one with the lowest initial price, but rather that which provides the most cost-effective approach to achieving long term goals. Initial cost must, of course, be a buying consideration, but it should not dominate the decision-making process.

It is also important to remember that the hardware and software of different manufacturers are not compatible. For this reason, once a Building Automation System is installed, the owner may very well be committed to continue with the same manufacturer for future extensions to the system. Consequently, it is extremely important to closely evaluate not only the system and its capabilities, but also the manufacturer's past experience in the Building Automation System business and his commitment to the future of that business.

Considerations in Selecting a Building Automation System

Differences between the actual capabilities of a BAS and the owner's original expectations can cause great disillusion and costly problems. To avoid this, the decision-making team must first establish a list of goals and objectives to be achieved by installing a BAS. Next, they should acquire a thorough understanding of BAS capabilities and how these apply to the stated goals and objectives. This is an educational process and should include technical, financial, and operational considerations, as well as potential future requirements for system expansion.

Existing mechanical system deficiencies should be reviewed and corrected prior to the installation of a BAS. If there are design problems with the heating, ventilating, and air-conditioning system (HVAC), not only will they not be corrected by the installation of a Building Automation System, but they may even be aggravated. Indoor air quality problems, hot spots, cold spots, humidity problems or unusually high energy costs could be caused by inadequate maintenance, poor control, or mechanical systems which are not properly installed or applied.

Economics play an important role in the decision-making process. Accuracy of control by DDC technology can result in significant savings over an obsolete pneumatic or electric control system. Energy management programs can generate impressive energy savings with little or no change in building comfort conditions. Integration of other building systems can also lead to operational savings through consolidation of building operations at a central location. Furthermore, installing a Building Automation System can result in more effective utilization of manpower. Therefore, it is important to do a complete financial analysis to determine the return on investment that can be achieved by the BAS.

Planning for future expansion of the BAS is another consideration. If building expansion or additional buildings are a future possibility, the BAS must have the capacity to meet the requirements associated with that growth.

Planning Phase

An experienced building operations person and an engineer familiar with Building Automation Systems are indispensable members of a BAS evaluation team. Their technical expertise and understanding of building operations can be invaluable in verifying potential savings to be achieved through energy management programs, and in avoiding the pitfalls of misapplication of the Building Automation System. They can also survey and evaluate existing building systems (HVAC, lighting, fire, security, and access control), and make recommendations to upgrade, retrofit, or replace equipment or complete systems where necessary. In the case of a new building, these individuals can work with the HVAC and electrical engineers to ensure that the BAS is properly incorporated into the building system's design.

Installation Phase

During the installation phase of the BAS, the owner will want a technical representative available to oversee the installation and review any job site revisions. Here again, the building operations person and the engineer with knowledge of BAS are the ideal candidates.

Formal Acceptance

Upon completion of the installation, a formal acceptance test should be conducted to verify that the system is functioning as specified. Every point should be tested, with the engineer and operations representative present. All software programs, DDC control loops, energy management, lighting fire, security and access control should be completely tested subject to the approval of the engineer. It is extremely important that the entire BAS perform in a manner that satisfies the goals and objectives established by the original decision-making team, before it is accepted.

Training

Prior to final acceptance of the BAS, a formal on-site training session should be conducted by the BAS vendor for all operating personnel. Training should include details of BAS operations as they apply to the specific building and its systems. Trainers should also review the goals and objectives of the decision-making team and explain how the BAS meets those requirements. Finally, all operators should receive in-depth training in their individual duties regarding proper operation of the system. The vendor should offer advanced operational and programming courses so that operators can advance through progressive levels of responsibility. This system provides incentives for the operators and also assures the building owner that his operations staff is fully qualified.

Maintenance Requirements

The maintenance requirements of computerized Building Automation Systems are minimal in comparison to pneumatic and electric control systems. However, vendor support in some form is necessary to ensure the goals and objectives of the team are met on an ongoing basis. While some facility managers are fortunate enough to have in-house technical expertise to maintain a BAS, this is the exception rather than the rule.

Vendor support offerings vary. Some vendors offer comprehensive maintenance programs at a fixed annual cost. Most vendors will tailor a maintenance program to the customer's needs. Some minimum services that should be considered by BAS system owners are covered in the following sections.

Software Support

Revisions and updates to standard operating software constitute an important BAS feature, the purpose of which is to avoid system obsolescence and enhance overall system operation. Some vendors conduct a periodic review of data and application programs to verify correct operation. This is extremely important for systems with fire and security integration.

Some vendors have remote diagnostic capability. This allows the vendor to access the BAS software from a remote location for trouble-shooting purposes. The remote operator can also change operating parameters, modify programs, or disable points.

Hardware Support

Since building automation and direct digital control systems are software-driven, hardware support is not as important as for traditional pneumatic and electric control systems. Software and applications support rank higher in importance than hardware support in BAS systems. From a budgetary standpoint, however, hardware support could be an important feature. Computer-related components and printed circuit boards can be expensive items. Under a service contract (which includes component replacement) an annual figure can be carried in the budget to cover unexpected replacement costs.

For systems with fire or life safety functions, hardware coverage could be critical. Most insurance companies require a fixed number of annual tests on initiating devices (fire detectors, smoke detectors, manual pull stations). Most vendors perform these required tests and furnish a written certification.

Emergency Service

Many vendors include emergency service on a seven-day, 24-hour basis as a feature of their service contracts. In fact, in many cases, a vendor will not respond to after-hours (overtime) emergency service calls unless the customer has a service contract specifically including this coverage. A system owner who has fire or life safety functions included in his system should include emergency service provisions. Some vendors will agree to a maximum response time for emergency calls if the building occupants or property will be adversely affected by a BAS malfunction. If the Building Automation System is critical to the health or safety of building occupants, or to the protection of property, a formal agreement should be made with the vendor for emergency service coverage.

Operator Training

Periodic operator training can be an important benefit to the BAS owner. Many vendors offer this feature to maintain a high level of operator proficiency. This type of training is usually conducted on-site, and the content determined jointly by the customer and the vendor. Some BAS manufacturers offer advanced formal training programs at their factories to upgrade operators in HVAC, fire, security and access control systems, building operations and BAS programming.

Vendor Selection Through Bidding Process

The decision-making team may decide that they wish to receive bids from multiple vendors for a Building Automation System. This complicates the decision-making process somewhat because of the different technology, experience levels, and conceptual approach offered by different vendors. For this reason, it is imperative that detailed specifications be produced to support the goals and objectives developed by the team. A vendor's representative may offer to provide off-the-shelf, non-proprietary specifications for the bid. This may be acceptable if the team reviews the documents carefully and customizes the wording where necessary. The consulting engineer and building operations person can provide invaluable assistance to the team in this task.

Evaluating Proposals

All proposals should be reviewed thoroughly and carefully by the decision-making team. The following list is presented as a guide to some of the key issues to consider in the areas of BAS functionality and vendor support.

- Central host computer malfunctions and breakdowns and their effect on the building systems.
- Costs and time required to alter the BAS data base and/or software.
- Costs of adding points after system is installed. User-friendliness to add and delete points and to change the software. Capability of performing changes without shutting down the system.
- Power failures and their effect on the BAS. System's capability to get back on-line after power is restored.
- Type and manufacturer of central host computer and whether it is a commonly used general purpose computer or specially configured for the BAS.
- Is the cost of software included with proposal or an added cost?
- How many password control levels of access?
- Access to future software revisions, enhancements, and programs.
- Central computer's capability to integrate DDC, energy management, lighting control, as well as fire and security/access and facility management programs.
- Automatic dial-up or dial-out communications to an off-site BAS or PC in the event of an abnormal condition.
- Capability to expand in order to control additional systems, buildings, and modifications to the original BAS.
- Automatic dial-up capability versus dedicated telephone line requirements to operate separate buildings.
- BAS application requirements for U.L. listing, such as lighting control, integrated fire and security/access systems.
- Experience and background levels of personnel who are to receive user training. Clarity and user-friendliness of the manufacturer's operator's manuals, and as-built documentation.
- Response time to repair a problem, stocking of repair parts, skill required.
- BAS maintenance requirements, availability, types, costs of manufacturer's maintenance agreements.
- Manufacturer and/or vendor experience in installing and servicing a BAS for the application.

The award should not be made based only on lowest price, but rather on *value*, as determined by evaluating the costs and benefits of each proposal. A common method used to evaluate proposals is value ranking. Each item of importance in the decision-making process is assigned a certain weight with an overall total, which may add up to 100 points. A value ranking system could be set up as shown in Figure 11.1.

Selecting a Vendor

After completing a proposal evaluation process, the selection should be narrowed down to two or three qualified vendors. In order to properly decide on the *best suited vendor*, the decision-making team should investigate and obtain as much information as possible about the vendor's past performance. At a minimum, some of the following actions should be taken:

- Site visit: View and inspect at least one, but preferably as many as possible, BAS's with similar applications.
- Discuss problems encountered by owners of similar BAS's.
- Interview the vendor's engineering, construction (if applicable), and maintenance staff who will be working on the project.
- Spend as much time as possible with the vendor in order to understand how the proposed BAS works, how user-friendly the software is, how to perform software programming, change start times, perform overrides, acknowledge alarms adding and deleting points, and any other item deemed important in the operation of the system.

The process of evaluating proposals and selecting a vendor requires some discipline and structure. The more extensive the review of the important factors, the more fundamentally sound the final vendor selection should be.

ITEMS	WEIGHTS
Proposal price	_____
Specification compliance	_____
Equipment quality based on proposal literature	_____
Hardware, software, and firmware quality based on proposal literature	_____
Expansion capability	_____
Fire system integration capability	_____
Security/access integration capability	_____
Lighting control integration capability	_____
System point capacity	_____
Number of installed systems and point size	_____
Options & features not specified	_____
Warranties	_____
Maintenance agreement costs	_____
Vendor experience	_____
Vendor financial stability	_____
Vendor reputation	_____
Engineering support & quality	_____
Field maintenance support & quality	_____
User training support & quality	_____
Installation quality	_____
Installation time schedule	_____
TOTAL	100

* Values to be determined by the decision-making team.

Example of a Value Ranking System

Figure 11.1

BAS Symbols

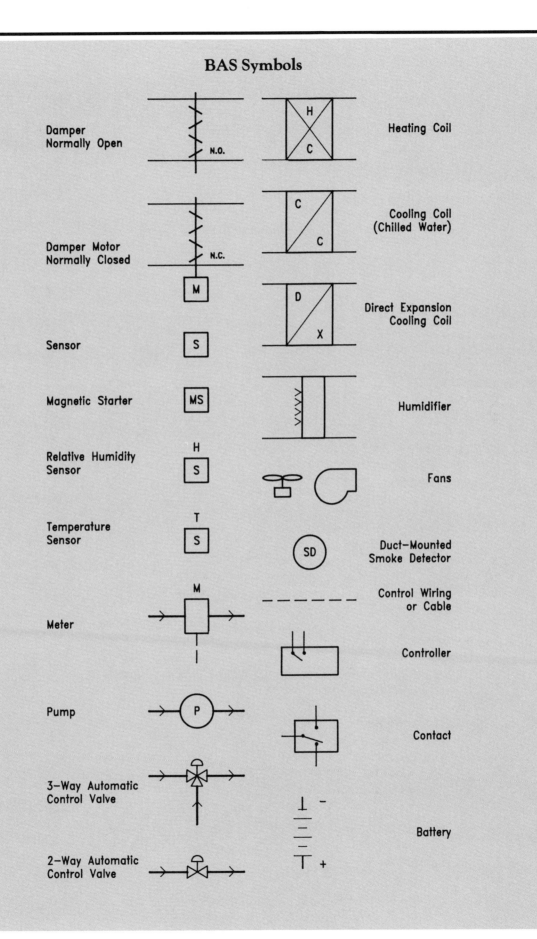

Damper Normally Open — N.O.

Damper Motor Normally Closed — N.C. — M

Sensor — S

Magnetic Starter — MS

Relative Humidity Sensor — H S

Temperature Sensor — T S

Meter — M

Pump — P

3-Way Automatic Control Valve

2-Way Automatic Control Valve

Heating Coil — H C

Cooling Coil (Chilled Water) — C C

Direct Expansion Cooling Coil — D X

Humidifier

Fans

Duct-Mounted Smoke Detector — SD

Control Wiring or Cable

Controller

Contact

Battery — + −

Glossary

Actuators
A device that converts an electric or pneumatic signal in order to position a control device such as a valve or damper.

AI
Analog inputs

Algorithms
A set of mathematical instructions, or a computer program, to produce a control output.

Analog
Continuously variable through a range. Not limited to discrete units.

AO
Analog outputs

Bell loop
The circuit that provides notification of a fire condition in a fire alarm system.

Bit
Abbreviation of "binary digit," the smallest unit of information in a binary system.

Built-up system
An HVAC system which is assembled on-site, usually a large system.

Central host computer
A personal computer (PC) that acts as an operator-machine interface to a Building Automation System.

Central Processing Unit (CPU)
The section of a computer where arithmetic and logical operations are performed, and instructions are interpreted and executed.

Centralized multiplexing system
A system of transmitting multiple signals simultaneously over a single circuit.

Cold deck (or duct)
Source of cold air, provides for cooling interior zones.

Communications bus

A communications medium which links controllers and a central host computer.

Control loop

A sensor-controller-actuator combination that processes an input signal and produces an output signal to position the actuator.

Direct expansion (DX)

Refrigeration systems that employ expansion valves or capillary tubes to meter liquid refrigerant into the evaporator.

Deadband program

See *zero-energy band program*

DDC

Direct Digital Controller

DI

Digital input

DO

Digital output

Digital transmission

The serial transmission of information, in bit form, over a single wire.

Direct Digital Controller (DDC)

A microprocessor-based device that controls and monitors HVAC equipment. Controllers typically have control loop programs and energy management programs.

Distributed process system

The consolidation of energy management and control loop programs into one intelligent microprocessor-based controller.

Dry bulb

The temperature as registered on a standard thermometer

Duct static pressure

The pressure acting on the walls of a duct; the total pressure less the velocity pressure; the pressure existing by virtue of the air density and its degree of compression.

Duty cycle

A program that turns electrical loads on and off while maintaining space comfort conditions, thereby conserving energy.

Economizer cycle

A system of dampers, temperature and humidity sensors, and actuators which maximizes the use of outdoor air for cooling.

Energy Management System (EMS)

A system which operates energy-consuming systems, in the most energy-efficient manner, through software programs.

Enthalpy program

A program which selects the source of air that contains the least total heat to be removed, thereby achieving savings on mechanical cooling.

Fan inlet vane damper

Adjustable blades or vanes at a fan inlet designed to give air a spinning motion in the direction of fan rotation.

Fire management panel

A fire alarm control unit consisting of multiple single zone fire alarm control modules, a telephone module, an HVAC module, and an audio communications module.

Firmware

Computer programs or instructions stored in the Read Only Memory (ROM). It performs the same function as software, but is in hardware form. Firmware is loaded into the equipment either at the time it is manufactured or later for preprogrammed controllers.

Flywheel effect

The tendency of a building to retain heat due to its mass-thermal-inertia.

Heat exchanger

A device designed to transfer heat between two physically separated fluids.

Hot deck (or duct)

Source of hot air, provides for heating interior zones.

Indicating circuit

See *bell loop*

Initiating circuit

A circuit containing sensors to detect a fire condition.

Integrated Building System

The consolidation of multiple building systems, such as HVAC, lighting, fire alarm, and security/access control into one central computer.

Intelligent (Buildings)

Building "intelligence" is defined by different levels of automation and integration of systems.

Kilobit

Thousand bits

Load reset program

A program that utilizes minimum levels of heating or cooling energy to satisfy the zone with the greatest heating/cooling requirement.

Load shedding

See *power demand program*.

Management host computer

A high level computer that acts as a management-machine interface to a BAS.

Microprocessor-based controllers

See *Direct Digital Controller*

Multi-conductor

An electrical cable containing multiple wires.

Night cycle program

A program that maintains a reduced temperature in the heating season or high limit temperature in the cooling season during unoccupied periods.

Night purge program
A program that replaces building air with cooler outdoor air during the early morning hours.

Off-normal alarms
An alarm generated by a deviation from preprogrammed limits.

Offset
A deviation between control point and set point on a modulating control loop.

Optimum start program
A program that delays HVAC system startup to the last possible moment, while still achieving comfort levels at occupancy time.

Optimum stop program
A program that takes advantage of the building flywheel effect to meet comfort requirements until the end of the occupancy period.

Pneumatic
A system of control that uses air pressure.

Point
A piece of equipment that is monitored or controlled by a controller or BAS.

Power demand program
The process of shedding electrical loads to avoid achieving a preprogrammed kilowatt demand limit.

Primary air
Conditioned air supplied to terminal units by a central HVAC system.

Protocol
A method by which microprocessors can communicate with each other in a common language.

Relay
An electromagnetic device activated by a variation in conditions in one electric circuit and controlling a larger current or actuating other devices in the same or another circuit.

Reset program
See *load reset program*

Sensing circuit
See *initiating circuit*

Set point
The setting of a controller (the desired level of control).

System controllers
A DDC controller used to control large central HVAC systems.

Time-based program
A program which starts and stops equipment as a function of preprogrammed time.

Variable Air Volume (VAV) system
An HVAC system which varies air flow to the conditioned space to maintain comfort conditions.

Variable speed drive
A method of controlling the fan speed in a central HVAC system.

VDC
Volts direct current

Wet bulb
The temperature read on a dry bulb thermometer enclosed in a wet wick.

Zero-Energy band program
A program that provides a control deadband between heating and cooling processes, during which no energy is consumed.

Zone controllers
A DDC controller used to control primarily terminal units or small package-type HVAC units.

Index

A

Access control systems, 140, (Fig. 8.3, p. 140). *See also* Security-access control systems.
Actuators. *See* Remote sensors and actuators.
Alarm summary, 94-95
All points summary, 94-95
Analog input (AI) point, 51, 53, (Fig. 3.4, p. 54), 15^
Analog input (AI) signal, 69, 70
Analog output (AO) point, 51, 54, (Fig. 3.5, p. 55), 159
Analog output (AO) signal, 70
Audio detectors, in security systems, 139
Automatic control systems
  electric, 15, 69
  electronic, 3, 12, 69
  microprocessor-based, 15, 69
  pneumatic, 3, 12, 15, 69

B

BAS. *See* Building Automation Systems.
Basic proportional control—cooling, (Fig. 2.2, p. 17)
Basic two-position control—heating (Fig. 2.2, p. 17)
Bell loop. *See* Indicating circuit.
Building Automation Systems and
  computerized maintenance management programs, 144
  and Direct Digital Control, (Fig. 4.26, p. 92)
  and energy management programs, 99-114
  and facility management functions, 143
  and fire safety integration, 121-136
  goals, 51
  integration, 62-65
  layout, 159-160, (Fig. 10.1, p. 161)
  and lighting control, 115-120
  and maintenance scheduling, 144-145
  monitoring tenant energy

  consumption, 149
  monitoring utility meters, 147, 149
  and security-access control integration, 137-142
  symbols, 185
  system selection criteria, 177, 179-184
Building Automation Systems,
  components, 10, 51-62, (Fig. 3.1, p. 52)
  central host computer (Level III), 51, 59, 61
  communications bus, 59
  management host computer (Level IV), 51, 61-62
  microprocessor-based controllers (Level II), 51, 56, 58-59
  remote sensors and actuators (Level I), 51, 53-54, 56
Building Automation Systems,
  development of, 3-13
  centralized control panels, 4-6
  centralized multiplexing systems, 7
  computerized systems, 7-9
  distributed process systems, 9-10
  early systems, 51
  future trends in, 11-12
  in response to design changes, 3
  local control panels, 3-4
Building Automation Systems, evaluation of a new system
  formal acceptance, 181
  installation phase, 181
  planning phase, 180-181
  training of operating personnel, 181
Building Automation Systems, monitoring efficiency of heating/cooling plants, 150-151, 153
  electric chiller plant, 153, (Fig. 9.14, p. 155), (Fig. 9.15, p. 156)
  fossil fuel hot water heating plant, 151, (Fig. 9.12, p. 153), (Fig. 9.13, p. 154)
  fossil fuel steam heating plant, 151, (Fig. 9.10, p. 151), (Fig. 9.11, p. 152)
Building management programs

action messages, 46
energy audit, 46, (Fig. 2.32, p. 46)
historical data, 45, (Fig. 2.31, 45)
material inventory, 46-47, (Fig. 2.33, p. 47)

C

Central host computer (Level III), 51, 59, 61, 62, 90-95
and building management programs, 45-47
color graphics/annunciation capabilities, 48, 94
components, 91, 110, 112
hardware, 59
CRT screen format (Fig. 4.27, p. 93)
dial in/out function, 50
and energy auditor program, 92-93
functions of, 15
group and historical logs, 94-95
line graphs and bar charts, 95, (Fig. 4.28, p. 96)
and maintenance management program, 47-48, 94
role in a distributed process system, 44-45
energy management programs, 110-112, 114
integrating fire safety systems, 134-136
lighting control systems, 119-120
security-access control, 142
software applications, 61, 92-95
storage and retrieval of historical data, 92
voice communications option, 48, 95, (Fig. 4.29, p. 97), (Fig. 6.4, p. 120)
Centralized control panels, (Fig. 1.2, p. 5)
development of, 4-5
graphic display of boiler and chiller plant, (Fig. 1.3b, p. 6)
graphic display of building floor plans, (Fig. 1.3a, p. 6)
limitations of early models, 5
benefits of early models, 5
Centralized multiplexing system, 7, (Fig. 1.4, p. 8)
system slide displays, (Fig. 1.5, p. 9)
Centrifugal chillers, 80. See also Cooling/dehumidification control, in DDC systems.
Communications bus, 10
defined, 59
types
direct-dial telephone lines, 59
fiber optic cable, 59
hardwire, 59
wiring configurations
daisy chain, 59, (Fig. 3.9, p. 60)
star, 59, (Fig. 3.10, p. 60)
Computer technology, and Building Automation Systems, 7-9
energy conservation, 8
hardware, 9
layout, (Fig. 1.6, p. 10)

minicomputers, 7
personal computers (PCs), 8
software packages, 7-8. See also Central host computer.
Contact devices, in security systems, 138
Control loops
defined, 16
in Direct Digital Control Systems, 15-16
in electric control systems, 15-16
in HVAC systems, 69
in pneumatic control systems, 15-16
Control panels, central. See Centralized control panels.
Control panels, local. See Local control panels.
Control systems. See Automatic control systems.
Control units
in fire safety systems, 127-129
fire management panels, 129
multiple-zone units, 129
single-zone units, 128-129
in security systems, 140
Cooling/dehumidification control, in DDC systems
chilled water control (centrifugal/reciprocating), 25, (Fig. 2.13, p. 26), 27, 80, (Fig. 4.13, p. 81) dehumidification control, 28 (Fig. 2.17, p. 29), 31
discharge air control (HVAC)— cooling, 27, (Fig. 2.14, p. 27)
discharge air reset control (HVAC)— cooling, 27-28, (Fig. 2.15, p. 28)
space temperature control—cooling, 28, (Fig. 2.16, p. 29)
Cooling plant efficiency. See Heating/cooling plant efficiency program.

D

DDC. See Direct Digital Control Systems.
DDC controllers, 16
and input devices, 17
analog, 17
digital, 18
standard software used in, 160
Deadband program, 83, 86, 88. See also Zero-energy band program.
Dehumidification control, 80, (Fig. 4.14, p. 82). See also Cooling/dehumidification Control.
Demand charges, 102-103
and building load profile, (Fig. 5.4, p. 103)
Demand interval, 102-103
Digital input (DI) point, 51, 53, (Fig. 3.2, p. 53), 159
Digital input (DI) signal, 69, 70
Digital output (DO) point, 51, 53, (Fig. 3.3, p. 54), 159
Digital output (DO) signal, 70
Direct Digital Control (DDC) Systems,

9-10, 12, 15-32, (Fig. 3.6, p. 55), 69-95, 157-158
  advantages over pneumatic and electric control systems, 16, 17
  applied to a typical HVAC system, 90, (Fig. 4.25, p. 91)
  and budget estimating, 166, (Fig. 10.5, pp. 174-175)
  control loops, 16
  controllers, 10, 12, 16-17
  converting a building to, 158
  cooling/dehumidification control in, 25, 27-28, 30, 77, 80
  employment of a 5-point control system, 54-56
  heating control in, 19-20, 22-23, 25, 73-75, 77
    constant temperature hot water control (boiler), 20
    constant temperature hot water control (heat exchanger), 20
    hot water reset control (boiler), 22
    hot water reset control (heat exchanger), 22
  discharge air control (HVAC)— heating, 23
  discharge air reset control (HVAC)— heating, 23
  heating-cooling sequencing in, 83, 90
  humidification control in, 30-31, 80
  humidification/dehumidification sequencing in, 84, 90
  input signals, 69
  and offset, 17
  output signals, 70
  role of central host computer in, 90-95
  sequence of operation, 161-163, 165-166
  static pressure control in, 31-32, (Fig. 2.21, p. 33), 84-85
  steps for constructing
    adding sensors and actuators, 160, (Fig. 10.2b, p. 163)
    entering the controller, 160, (Fig. 10.2c, p. 164)
    point summary, 160, (Fig. 10.3, p. 165)
    schematic layout, 160, (Fig. 10.2a, p. 162)
  variable air volume (VAV) terminal units, 85-86, 88-89, 165-166
  ventilation control in, 18-19
    economizer control of mixed air, 19
    fixed quantity of outdoor air, 18
    mixed air control, 18
Direct expansion (DX) cooling, 27, 28
Distributed process systems, 15-50, (Fig.2.1, p. 16)
  and building security-access control, 44
  and Direct Digital Control, 9-10, 15-32
  and energy management programs
    HVAC, 32-41
    lighting, 41-42
  and fire safety systems, 43-44
  layout, (Fig. 1.7, p. 11)
  main components, 44-45
  role of central host computer in, 44-45
Duty cycle program, 100-101

and effect on electric motors, 101
  guidelines for application, 101. See also Energy management programs.
  schedule load levelling, (Fig. 5.2, p. 101)
  space temperature compensated, (Fig. 5.1, p. 100)
DX cooling. See Direct expansion cooling.

E

Electric control systems, 15-16
  components, 15
  and offset, 17
Electrical consumption
  demand charges, 102-103
  demand interval, 102-103
Electronic control systems, 3, 12
EMS. See Energy Management Systems.
Energy audit program, 92-93, 112. See also Building management programs.
Energy conservation, and computerized BAS, 8
Energy consumption
  breakdown by system, 99
  monitoring. See Utilities metering program; see tenant energy monitoring program.
  sources of savings, 99
Energy management programs, 99-114, 160, 165
  duty cycle program, 32-33, (Fig. 2.22, p. 34), 100-101
    fixed time duty cycling, 32-33
    variable time duty cycling, 32-33
  enthalpy program, 36, (Fig. 2.27, p. 39), 107-108
  lighting level control, 42, (Fig. 2.30, p. 42)
  load reset program, 36, 39, (Fig. 2.28, p. 40), 40, 108-110
  occupied-unoccupied program (lighting systems), 41-42
  optimum start/stop program, 35-36, (Fig. 2.24, p. 37), 105, 107
  power demand limiting program, 33-34, (Fig. 2.23, p. 35), 102-104
  software used to evaluate, 112-114, (Fig. 5.11, p. 113)
  time of day program, 35
  unoccupied night purge program, 36, (Fig. 2.26, p. 38), 107
  unoccupied period programs, 104-105
  unoccupied space temperature setback (night cycle) program, 36, (Fig. 2.25, p. 38)
  zero-energy band program, 40-41, (Fig. 2.29, p. 41), 110
Energy Management Systems (EMS), 59
Energy savings
  sources of, 99
Enthalpy program, 107-108

F

Facility management programs, 143-153
  heating/cooling plant efficiency program, 150-151, 153
  maintenance, 144-145

tenant energy monitoring program, 149
utilities metering program, 147, 149
Fire alarm control units. *See* Control
units, in fire safety systems.
Fire detection sensors. *See* Initiating
circuits, in fire safety systems.
Fire management panels, 129, (Fig. 7.15,
p. 131)
Fire safety integration, 121-136. *See also*
Fire safety systems.
Fire safety systems, 43-44
annunciation and alarm, 43-44
components, 122-129, (Fig. 7.2, p. 122)
control units, 122, 127-129
indicating circuits, 122, 126-127
initiating circuits, 122-126
detection methods, 43
function of, 121
integration through BAS, 43, 132-134
partial, 132, (Fig. 7.16, p. 133)
total, 132-134, (Fig. 7.17, p. 134)
role of central host computer in, 134-136
smoke pressurization control, (Fig. 7.18,
p. 135)
stand-alone, 43, 130-131
types, 121, (Fig. 7.1, p. 122)

G
Group point summary, 94-95

H
Heating control, in DDC systems, 19-25
and ventilation, 70, 72-73
constant temperature hot water control
boiler, 20, (Fig. 2.6, p. 21), 22, 73,
(Fig. 4.4, p. 74), (Fig. 4.5, p. 75)
heat exchanger, 73, (Fig. 4.6, p. 76),
20, (Fig. 2.7, p. 21), 22
discharge air control (HVAC), 23, (Fig.
2.10, p. 24), 74-75, (Fig. 4.10, 78)
discharge air reset control (HVAC), 23,
(Fig. 2.11, p. 24), 75, (Fig. 4.11, p. 79)
hot water reset control
boiler, 22, (Fig. 2.8, p. 22), 73, (Fig.
4.7, p. 76), (Fig. 4.8, p. 77)
heat exchanger, 22, (Fig. 2.9, p. 23),
74, (Fig. 4.9, p. 78)
space temperature control (HVAC), 25,
(Fig. 2.12, p. 25), 75, 77, (Fig. 4.12, p. 79)
Heating/cooling plant efficiency
program, 150-151, 153
Historical trend logs, 95
Humidification control, in DDC Systems,
(Fig. 2.18, 30), 30-31, 80, (Fig. 4.15, p. 82)
heating/cooling sequencing, 30, (Fig.
2.19, p. 31)
humidification/dehumidification
sequencing, 31, (Fig. 2.20, p. 32)
HVAC
application of DDC system to, 90

I
Indicating circuit devices
in fire safety systems, 126-127
bells, (Fig. 7.11, p. 127)

horns, (Fig. 7.10, p. 127)
signal with strobe, (Fig. 7.12, p. 128)
in security systems, 139
Induction motors, guidelines for
cycling, 101, (Fig. 5.3, p. 101)
Infrared detectors, in security
systems, 139
Initiating circuits
in fire safety systems, 122-126
manual pull stations, 122, 125-126
smoke detectors, 122, 123-124
sprinkler heads (water flow
sensors), 122, 124-125
thermal detectors, 122, 123
in security systems, 138-139
audio detectors, 139
contact devices, 138
infrared detectors, 139
Input devices
in DDC controllers, 17-18
Integration, of Building Automation
Systems, 10-11, 12-13, 62-65
barriers to, 13
effect of microelectronic technology on, 15
layout, (Fig. 1.8, p. 12)
methods, 62-65, (Fig. 3.12, p. 63), (Fig.
3.13, p. 64)
role of central host computer in, 15
Ionization (smoke) detectors, 123-124,
(Fig. 7.5, p. 125)

L
Life safety systems. *See* Fire safety systems,
types.
Lighting control
categories of, 41
in a building automation system, 115-120
steps to reduce costs, 115
building requirements, 115
strategies, 116-118
occupied-unoccupied lighting
control, 116
lighting level control, 116-118. *See
also* Energy management programs
(lighting).
Load reset program, 108-110
and discharge air reset (Fig. 5.9, p. 109).
See also Energy management programs.
Local control panels, 3-4, (Fig. 1.1, p. 4)

M
Maintenance management programs, 94
computerized, 47-48, (Fig. 2.34, p. 49), 144
financial analysis, 145, (Fig. 9.7, p. 148),
(Fig. 9.8, p. 149)
maintenance history, 144
maintenance scheduling, 144-145
management information reports, 145
material inventory, 144, (Fig. 9.5, p.
147), (Fig. 9.6, p. 148)
work order printout, 144, (Fig. 9.3, p.
146), (Fig. 9.4, p. 146)
Maintenance requirements and vendor
support
Maintenance scheduling, 144-145, (Fig.

9.1, p. 145), (Fig. 9.2, p. 145)
Management host computer (Level III), 51, 61-62
Manual pull stations, 125-126, (Fig. 7.9, p. 126)
Material inventory programs. *See* Building management programs; see Maintenance management programs.
Means Mechanical Cost Data, 158, 166, (Fig. 10.4, pp. 167-173)
Microprocessor-based controllers (Level II), 51, 56, (Fig. 3.7, p. 57), 58-59
  types
    zone controllers, 56, 58
    system controllers, 58, 59
  vs. traditional control systems, 56
Microprocessor-based fire alarm systems, 130-131
Mixed air economizer control system, 107
Modulated lighting control, 116, 117-118, (Fig. 6.3, p. 119)
  vs. standard lighting levels and power consumption, (Fig. 6.2, p. 118)
Multi-level lighting control, 116, (Fig. 6.1, p. 117)
Multiple-zone fire alarm control unit, 128-129, (Fig. 7.14, p. 130)
Multiplexing systems. *See* Centralized multiplexing system.

N

National Fire Protection Association (NFPA), 43, 122
NFPA. *See* National Fire Protection Association.
Night cycle program. *See* Unoccupied space temperature setback program.
Night purge program. *See* Unoccupied night purge program.

O

Offset, 17
  and Direct Digital Control, 17
  and electric control, 17
  and pneumatic control, 17
Optimum start/stop program, 105, (Fig. 5.6, p. 106), (Fig. 5.7, p. 106), 107. *See also* Energy management programs.
Output devices
  in DDC controllers, 17-18

P

Personal computers, and BAS, 9-10
  hardware, 9
Photoelectric (smoke) detector, 124, (Fig. 7.6, p. 125)
Pneumatic control systems, 3, 12, 15, 157-158
  components, 15
  and offset, 17
Pneumatic control valve (AO) device, (Fig. 3.5, p. 55)
Points, 51, 53-54, 56, 159
  analog input (AI), 51, 53
  analog output (AO), 51, 54

digital input (DI), 51, 53
digital output (DO), 51, 53
example of a 5-point control system, (Fig. 3.6, p. 55)
sequence of operation, 54, 56
Power demand limiting program, 102-104
  and load prioritizing, 103-104, (Fig. 5.5, p. 104). *See also* Energy management programs.
Program clocks, 105
Property protection systems. *See* Fire safety systems, types.

R

Ratchet clause, in establishing demand charges, 103
Reciprocating chillers, 80, (Fig. 4.13, p. 81). *See also* Cooling/Dehumidification control, in DDC Systems.
Remote sensors and actuators (Level I), 51, 53-54, 56. *See also* Points.

S

SCR. *See* Silicon Control Rectifier.
Security-access control systems, 44, 137-140
  alarm circuit, (Fig. 8.2, p. 139)
  categories of protection, 137-138
  components, 138-140, (Fig. 8.1, p. 139)
  control unit, 140
  functions of, 138
  indicating circuit devices, 139
  initiating circuit devices, 138
  integration, 141-142 (Fig. 8.4, p. 141)
Sensing circuit. *See* Initiating circuits.
Sensing loop. *See* Initiating circuits.
Sequencing, in DDC Systems, 83-84, 90
  heating-cooling, (Fig. 4.16, p. 83), 90
  humidification/dehumidification, (Fig. 4.17, p. 84), 90
Silicon Control Rectifier, 23, 75
Single-zone fire alarm control units, 128-129, (Fig. 7.13, p. 129)
Smoke detectors, types
  ceiling-mounted, 123-124
  duct-mounted, 124, (Fig. 7.7, p. 125)
Space temperature control (HVAC)— heating, 25
Sprinkler heads, 124-125, (Fig. 7.8, p. 126)
Stand-alone fire safety systems, 130-131
  fire alarm control unit, 130
  fire management system, 131
  intelligent fire alarm system, 131
Static pressure control, in DDC Systems, 31-32, 84-85, (Fig. 4.18, p. 85)
Static pressure sensor (DI), (Fig. 3.2, p. 53)
System controllers, (Fig. 3.8, p. 57), 58, 59

T

Tenant energy monitoring program, 149, (Fig. 9.9, p. 150)
Thermal detectors, 123
  fixed temperature, 123, (Fig. 7.3, p. 123)

rate-of-rise/fixed temperature, 123, (Fig. 7.4, p. 124)

Time of day program. *See* Energy management programs.

U

Unoccupied night purge program, 107, (Fig. 5.8, p. 108). *See also* Energy management programs.

Unoccupied period (night cycle) program, 104-105

Unoccupied space temperature setback (night cycle) program. *See* Energy management programs.

Utilities metering program, 147, 149

V

Variable air volume (VAV) system
  terminal box control applications, 85-86, 88-89
    cooling and electric reheat, 86, (Fig. 4.20, p. 87)
    cooling and hot water reheat, 86, (Fig. 4.21, p. 87)
    fan-powered return air and electric heat, 88, (Fig. 4.23, p. 89)
    fan-powered return air and hot water heat, 88, (Fig. 4.24, p. 89)
    fan-powered return air heat, 86, (Fig. 4.22, p. 88)
    night temperature setback-setup/morning warm-up-cool-down, 88, 90
    pressure independent—cooling only, 86, (Fig. 4.19, p. 86)

Vendor selection, 182-184
  evaluating bid proposals, 183
  value ranking, 183, (Fig. 11.1, p. 184)
  investigating past performance, 183-184

Vendor support
  emergency service, 182
  hardware support, 182
  operator training, 182
  software support, 181

Ventilation control, in DDC systems, 18-19, 70, 72-73
  economizer control of mixed air, 19, (Figure 2.5, p. 20), 72-73, (Fig. 4.3, p. 72), 90
  fixed quantity of outdoor air, 18, (Figure 2.3, p. 18), 70, (Fig. 4.1, p. 71)
  mixed air control, 18, (Figure 2.4, p. 19), 70, (Fig. 4.2, p. 71)

W

Water flow sensors. *See* Sprinkler heads.

Z

Zero-energy band program, 110, (Fig. 5.10, p. 111). *See also* Energy management program.

Zone controllers, 56, 58, (Fig. 3.8, p. 57)
  typical point data, 58